5th International Conference on Energy Materials and Applications (ICEMA 2020)

Held online due to COVID-19

IOP Conference Series: Materials Science and Engineering
Volume 908

Paris, France
6 - 9 May 2020

ISBN: 978-1-7138-3164-8
ISSN: 1757-8981

Printed from e-media with permission by:

Curran Associates, Inc.
57 Morehouse Lane
Red Hook, NY 12571

Some format issues inherent in the e-media version may also appear in this print version.

Additional copies of this publication are available from:

Curran Associates, Inc.
57 Morehouse Lane
Red Hook, NY 12571 USA
Phone: 845-758-0400
Fax: 845-758-2633
Email: curran@proceedings.com
Web: www.proceedings.com

TABLE OF CONTENTS

IOP Conf. Series: Materials Science and Engineering **908** (2020) 011001 doi:10.1088/1757-899X/908/1/011001

PREFACE

ICEMA providing a yearly platform for delegates and members to present and discuss the latest research, and our delegates and members will have many opportunities engage in dialogues about Energy Materials and Applications. It also provides new insights and bring together scholars, scientists, engineers and students from universities and industry all over the world under one roof.

This year, ICEMA was held online to ensure authors and speakers can be away from COVID-19 and keep safe. The conference was held virtually for 4 days including the online test at the first day, 2 keynote speeches and 1 authors' presentation session on May 7th, 3 invited speeches as well as 1 authors' presentation session on June 8th. On June 9th, we made the recording display for those who can't attend the conference because of the internet connections problem. The Conference has been a good opportunity for participants coming from Japan, Germany, Mexico and so many countries to present and discuss topics in their respective research areas.

The conference program was highlighted by our keynote speeches delivered by renowned scholars who shared their new research perspectives with our conference attendees. They are Prof. Lars Johanning from University of Exeter, UK, Prof. Jesus Toribioo from University of Salamanca, Spain, Prof. Monica Siroux from INSA Strasbourg ICUBE, University of Strasbourg, France, Prof. Jun Peng from University of Bedfordshire, Luton, UK. Prof. Nezihe AYAS from Eskisehir Technical University, Turkey. The authors' presentations cover topics on New Energy Generation Technology and Power System, Energy Materials and Environmental Science. Each author has 15 minutes (including Q&A). The outstanding performance of all the presenters made this conference truly spectacular.

The remote conference brings an interesting and convenient experience to ICEMA2020. However, due to the limitations of time zones and network, there are still some regrettable. We are looking forward to more direct communication between participants in the next year.

We would like to express our sincere appreciation to Conference Chairs, Program Chairs, Technical Committee Members, and international reviewers. Their excellent professional advice enabled us to prepare this conference. Our special thanks go to our speakers as well as all the authors for sharing their latest research results. We truly believe that the participants have had fruitful discussions, and enjoyed the opportunity for future collaborations.

We are looking forward to welcoming you next year.

Best regards,

Conference Chair
Prof. Jesus Toribio
University of Salamanca, Spain

IOP Conf. Series: Materials Science and Engineering **908** (2020) 011002 doi:10.1088/1757-899X/908/1/011002

COMMITTEES

Conference Chair

Prof. Jesus Toribio, University of Salamanca, Spain
Prof. David Lin, University of Tainan

Program Chairs

Prof. Jun Peng, University of Bedfordshire, Luton, UK
Prof. Nezihe Ayas, Eskisehir Technical University, Turkey
Prof. Monica Siroux, INSA Strasbourg, France

Technical Committee

Adhikari Rajendra Singh, Politecnico di Milano, Italy
Andoni Lasheras Aransay, Universidad del país vasco (UPV/EHU), Spain
B. Narayanamoorthy, Jiangsu University, China
Chandra Shekhar Malvi, Madhav Institute of Technology & Science, India
Chun-Ping Jen, National Chung Cheng University, Taiwan
Dai-Viet N. Vo, Universiti Malaysia Pahang, Malaysia
Francesca Scalisi, University of Palermo, Italy
Francisco Lopez-Huerta, Centro de Investigación en Micro y Nanotecnología, Mexico
Haryati Yaacob, UTM, Malaysia
Hung-Pin Hsu, Ming Chi University of Technology, Taiwan
Ipek Guler, Cankaya University, Turkey
Kien Wen Sun, National Chiao Tung University, Taiwan
Matteo Gentilucci, University of Camerino, Italy
Miguel Lorenzo, University of Salamanca, Spain
Najma Laaroussi, Mohammed V University of Rabat, EST de Salé, Morocco
Ngoc-Chung Le, Da Lat University, Viet Nam
S. Harish, Kyushu University, Japan
Vijay Kumar Gupta, PDPM Indian Institute of Information Technology Design and Manufacturing, India
Wegdan Wagdy El-Nadoury, Pharos University, Egypt
Yanan Hao, Beijing University of Posts and Telecommunications, China

Peer review declaration

All papers published in this volume of *IOP Conference Series: Materials Science and Engineering* have been peer reviewed through processes administered by the Editors. Reviews were conducted by expert referees to the professional and scientific standards expected of a proceedings journal published by IOP Publishing.

- **Type of peer review:** Double-blind
- **Conference submission management system:** Iconf submission system
- **Number of submissions received:** 22 papers
- **Number of submissions sent for review:** 22 papers
- **Number of submissions accepted:** 7 papers
- **Acceptance Rate (Number of Submissions Accepted / Number of Submissions Received X 100):** 31.82%
- **Average number of reviews per paper:** 4 papers
- **Total number of reviewers involved:** 12 reviewers
- **Any additional info on review process:**
  1. Preliminary Review
  2. The papers passed the first review will be eviewed again from the following aspects: Originality, Innovation, Technical Soundness, Applicability, Clarity of presentation and Relevance

Contact person for queries: Tatiana Fan **(**tatiana_fan@young.ac.cn)

IOP Conf. Series: Materials Science and Engineering **908** (2020) 012001 doi:10.1088/1757-899X/908/1/012001

Light-induced In-situ Ti^{3+} Formation in TiO_2 Nanosheets for Photocatalytic Hydrogen Evolution

E Wierzbicka[1], B Osuagwu[1], N Denisov[1], D Fehn[2], K Meyer[2] and P Schmuki[1]

1 Department of Materials Science, WW4-LKO, University of Erlangen-Nuremberg, Martensstraße 7, 91058, Erlangen, Germany
2 Department of Chemistry and Pharmacy, Inorganic Chemistry, Friedrich-Alexander University Erlangen-Nürnberg (FAU), Egerlandstrasse 1, 91058 Erlangen, Germany.
E-mail: ewa.wierzbicka@fau.de

Abstract. TiO_2 has been the benchmark semiconductor for the production of photocatalytic H_2 from aqueous media (with and without sacrificial agent). On TiO_2 surfaces, the photocatalytic H_2 evolution reaction in aqueous environments is kinetically severely hampered. To overcome this limitation and reach reasonable H_2 generation rates, a well-elucidated approach is the use of noble metal co-catalysts. In contrast to costly noble metal approaches, it recently has been reported that titania reduction treatments can lead to a noble-metal-free photocatalytic H_2 generation. So-called "grey" titania due to Ti^{3+} states shows intrinsically activated photocatalytic H_2 evolution [1–3]. The present work demonstrates the feasibility to use *in-situ* photoinduced reduction to create Ti^{3+} states that act as intrinsic catalyst and activate hydrothermal synthetized anatase nanosheets for H_2 generation to mediate the transfer of photo-induced charge carriers to the electrolyte.

1. Introduction

Since the discovery of photocatalytic water splitting by Fujishima and Honda in 1972 [4], titanium dioxide has become the benchmark semiconductor for the photocatalytic production of H_2 from aqueous media (with and without sacrificial agents) [5–7]. The TiO_2 conduction band position provides the fundamental energetic precondition to produce hydrogen from pure water or alcoholic aqueous solutions, however this reaction is kinetically severely hampered [5,6,8]. Accordingly, many efforts have been made to improve the photocatalytic activity of TiO_2. The outstanding efficiency was achieved by noble metal co-catalysts deposition on TiO_2 [5,6,9,10], however the most frequently these co-catalysts are costly noble metals such as Pt, Pd or Au. More recently alternative approach has been reported that an intrinsic defect formation in titania can improve photocatalytic hydrogen generation. So-called "grey" titania obtained from pure TiO_2 by simple hydrogen annealing at elevated temperatures shows significant activation for photocatalytic H_2 evolution [3,11–13]. An observed coloration of the material is ascribed to surface defects formation (Ti^{3+}-O_v). Those defect states are able to mediate photoelectron transfer to liquid phase. Another remarkable recently reported observation is that reduced states in titania (Ti^{3+}-O_v) can be simply induced by illumination in oxygen-free water-methanol solution [14]. This finding is particularly interesting due to the simplicity of "one-pot" approach, where light firstly triggers the formation of co-catalytic defect states that then increasingly cause an enhanced H_2 evolution. Chen and Mao who first described reduced 'black' titania ascribed higher photocatalytic performance to the formation of an amorphous layer at the outer part of TiO_2 nanoparticles [15]. In our previous study on light induced formation of Ti^{3+}-O_v, the presence of amorphous layer was unclear due to high roughness of commercial anatase particles [14]. Considering above theories, herein, we investigate the photoinduced reduction to create in-situ Ti^{3+} states in hydrothermally synthetized TiO_2 nanosheets under

IOP Conf. Series: Materials Science and Engineering **908** (2020) 012001 doi:10.1088/1757-899X/908/1/012001

long time illumination (6 days). Applied hydrothermal synthesis conditions allow the formation of crystalline anatase with very well defined particles shape and sharp edges which made it possible to clearly verify if there is any change in TiO_2 surface state morphology induced by light.

2. Methodology

Anatase crystalline powder with nanosheets morphology were synthesized by a simple hydrothermal method using commercially available reagents: Tetrabutyl titanate ($Ti(OBu)_4$, Sigma Aldrich) and concentrated hydrofluoric acid (48% HF, Merck). 1.2 ml HF was added dropwise to 10 ml of $Ti(OBu)_4$) placed in a 250 ml Teflon liner, under stirring at RT. The stirring was continued for the next 40 minutes before the Teflon containing the mixture was sealed in an autoclave which was eventually transferred to a preheated oven at 200 °C. The reaction was completed after 24 h and the autoclave was allowed to cool in the oven. After hydrothermal reaction, the precipitates were collected and washed with distilled water and ethanol several times and finally dried in oven at 75 °C overnight.

The morphologies of the as-synthesized powder was characterized by a field emission scanning electron microscope (Hitachi S4800). Average particles had a size of 20-60 nm with mainly exposed [001] facets. Crystal structure of synthetized powder was investigated by XRD using an X-ray diffractometer (X'pert Philips PMD diffractometer) with a Panalytical X'celerator detector and CuKα radiation (λ=1.54056 Å). For photocatalytic H_2 evolution measurements under open circuit condition, 10 ml suspension of 2 mg powder in oxygen-free (by 30 minutes N_2 purging) water - methanol (1:1) solution was prepare in a sealed quartz tube. The suspension was continuously illuminated under a 365 nm LED (175 mW/cm^2). A gas chromatograph (GCMS-QO2010SE, SHIMADZU) with TCD detector was applied to evaluate the concentration of produced H_2. In the first cycle (24 h) of experiment, sampling was taken before and after 1. 3, 6, 9 and 24 h hours of illumination from the gas in the headspace of quartz tube. To evaluate long-term maintenance of photocatalytic activity of the powder, such cycles were repeated three times (3 x 24 h), separated by 30 minutes purge with N_2 to remove previously formed H_2 from the system.

Electron Paramagnetic Resonance (EPR) spectra were recorded on a JEOL continuous wave spectrometer JES-FA200 equipped with an Xband Gunn diode oscillator bridge, a cylindric mode cavity and a N_2 cryostat. The samples were measured in the solid state under argon atmosphere in quartz glass EPR tubes at 95 K with a similar loading of ≈ 20 mg. The EPR quartz tubes with samples and electrolyte were irradiated under LED, 365 nm, 175 mW/cm^{-2} for different time intervals. The spectra were measured with the following parameters: Temperature 95 K, microwave frequency ν = 8.959 GHz, modulation width 1.0 mT, microwave power 1.0 mW, modulation frequency 100 kHz and a time constant of 0.1 s. Data analysis and simulation was carried out using the software "eview" and "esim" written by E. Bill (MPI for Chemical Energy Conversion, Mülheim an der Ruhr).

An X-ray photoelectron spectrometer (XPS, PHI 5600 XPS spectrometer, US) was used for compositional analysis of the samples. All XPS spectra were shifted to C1s peak at 284.8 eV. All samples were transferred from reaction quartz tube to the XPS stage in an argon atmosphere. The analysis of XPS spectra was performed using Multipack software.

Transmission electron microscopy was carried out with a Philips CM30 TEM. Geometrical features were measured from TEM images using Image J software.

3. Results and discussion

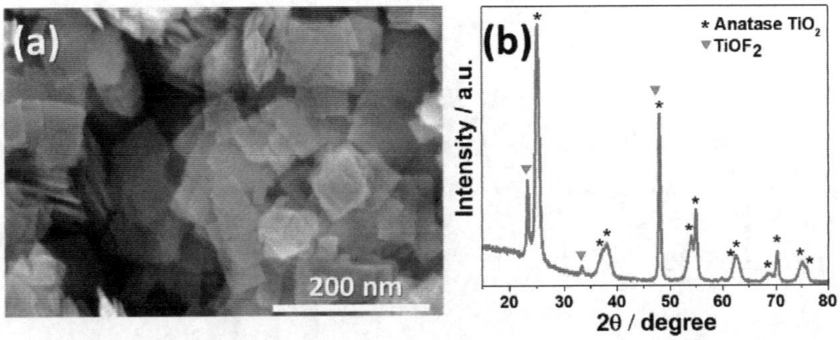

Figure 1. (a) SEM images of as received nanosheets (powder form). (b) XRD patterns of as received powder.

Figure 1a shows the SEM images of the powder used in this study before illumination. Optimized hydrothermal synthesis conditions allow to obtain powder of nanosheets form with very well defined particles shape, sharp edges and size between 20 and 60 nm. This method of synthesis enabled morphological control to expose specific crystal facets. As received nanosheets already have crystallographic structure of anatase TiO_2 confirmed by XRD (Figure 2b), with characteristic diffraction pattern of anatase (PDF card No. 00-021-1272) and the peaks position at 25.3°, 37.8°, 38.6°, 48.0°, 53.9°, 55.1° ,62.7°, 68.8°, 70.3°, 75.0° and 76.0°. The peaks placed at 23.4°, 33.4° and 47.9° correspond to titanium oxydifluoride which is impurity from hydrothermal synthesis reaction.
The cyclic photocatalytic hydrogen evolution performance from anatase nanosheets powder under a UV light source (LED, 365 nm, 175 mW/cm^{-2}) is presented in Figure 2a. Similar to our previous studies cycling experiments show that no H_2 evolution could be detected for the first hours due to a time prerequisites for initiation of Ti^{3+}-O_v active centers. The H_2 generation have characteristic exponential growth only in the first cycle (first 24 h) [14]. Despite of illumination interruptions and nitrogen purging between the cycles, light-induced Ti^{3+}-O_v sites remain active in the following cycles. A steady, nearly linear increase of the H_2 production is observed. The average H_2 evolution rate reaches about 125 µmol h^{-1} g^{-1} in every cycle, during 3 times repeated cycles (3 x 24 h) of illumination. Notably, a change of color of the nanopowders samples can be observed during UV light exposure as shown in Figure 2b. Powders darkening is a result of light-induced formation of Ti^{3+}-O_v surface defects. It was previously shown in literature that in absence of oxygen in the reaction environment, reduction leads to the formation of color Ti^{3+} states [13,14,16]. This Ti^{3+} defects can act as an intrinsic co-catalyst for photocatalytic H_2 evolution due to their introduction create additional electronic states below the conduction band [16].
Formation of those active centers was confirmed by EPR spectroscopy (Figure 2c). For these measurements, nanosheets anatase powder slurries in O_2-free 50% methanol-waters solutions were placed in quartz EPR tubes and exposed to different illumination time with LED 365nm (175 mW/cm^2) up to 6 days (Figure 2b). The measurements of pristine TiO_2 powder (Figure 2c inset, 0 h) show a low-intensity, sharp EPR signal from naturally present Ti^{3+} defects in regular lattice position centered at $g_x = g_y = 1.99$ and one $g_z = 1.96$. Clearly, illumination (short time 3h and long time 6 days) of the nanopowders gives rise to a strong paramagnetic signature centered at a g-value of $g = 1.94$ [17] (Figure 2c) that could be fitted by the resonance of Ti^{3+} surface exposed species. Note that the signal for 6 days illumination is 10 times attenuated.

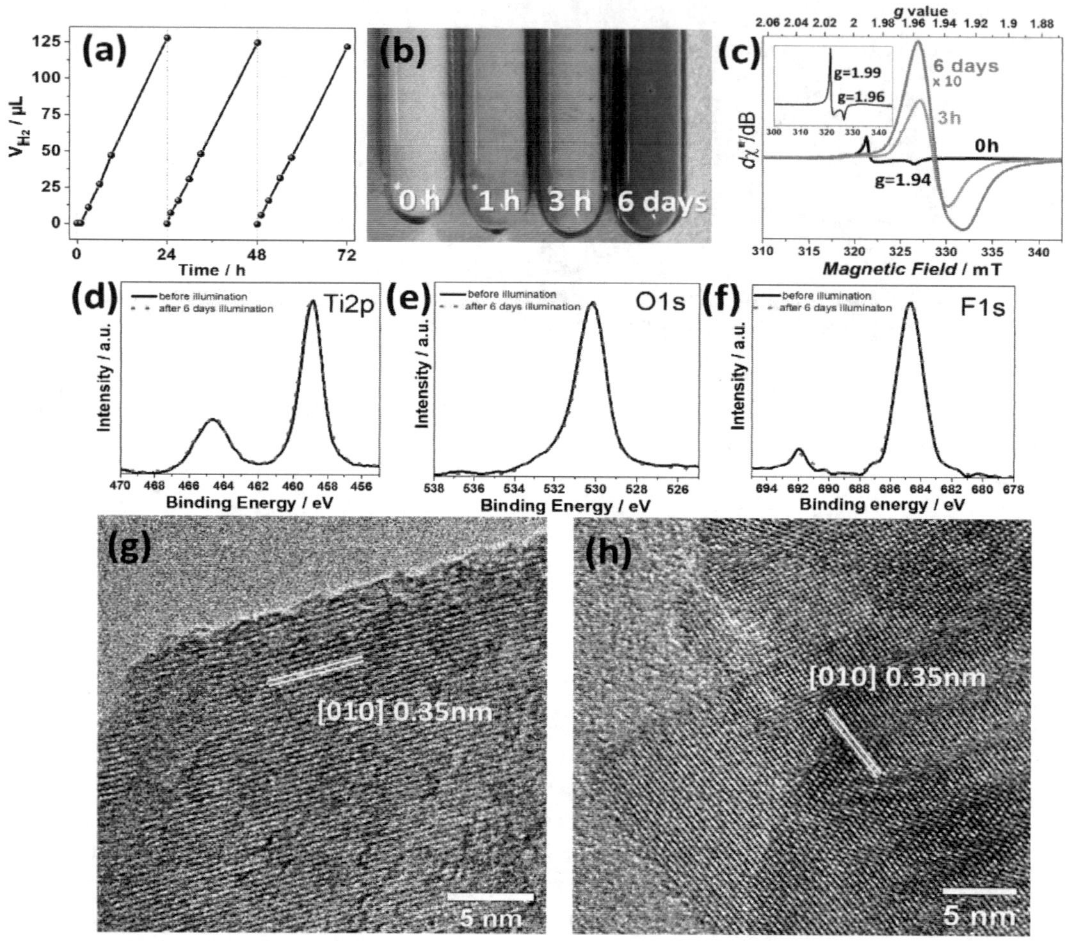

Figure 2. (a) Volume of generated H_2 on 2 mg of anatase powder (nanosheets structure) in 10 mL H_2O: methanol (50:50 vol.%) in cycling experiment under UV 365 nm light. (b) Photography illustrating changes of anatase powder (nanosheets structure) color after varied time of illumination with UV 365 nm light: (from the left side) 0 h, 1 h, 3 h and 6 days. (c) EPR spectra of the powder after varied time of illumination with UV 365 nm for 0 h, 3 h and 6 days (please note that the intensity of the sample after 6 days illumination is attenuated by a factor of 10). (d,e,f) XPS data of the powder before and after UV illumination. HR-TEM images of the nanosheets (g) before and (h) after light treatment.

The above experimental findings therefore clearly point toward a Ti^{3+}-O_v sites defect formation introduced by the illumination in anatase powder with nanosheet morphology.

The sample before and after exposure to UV light for 6 days were further studied by XPS. In the Ti2p, O1s and F1s spectra of the powders before and after illumination no changes are apparent (Figure 2d,e,f) which indicates that the concentration of Ti^{3+} states are below the XPS detection limit. Overall, HR-TEM images of the particles g) before and h) after 6 days UV illumination does not show difference in morphology (lattice fringes are evidently intact). We have identified that even long time illumination with strong UV light source (175 mW/cm^{-2}) that induced Ti^{3+}-O_v states formation [14], do not lead to amorphous shell formation. This result is in contradiction to the generally accepted theory claims that amorphous shell (large concentration of Ti^{3+}-O_v states) is necessary to form an additional mid-gap states overlapping with the conduction band edge which promote H_2 evolution, while small defect concentration leads to rapid recombination of photogenerated charge carriers. It has been proven many times that the defect sites in the bulk of photocatalysts introduced by doping can result with recombination centres formation [18]. Our finding show that surface-localized defects like Ti^{3+}-O_v, even trace amounts, acts as metal-like co-catalyst for enhanced photocatalytic H_2 generation.

4. Conclusions
In summary, the present study demonstrates an activation of nanosheets form anatase powder for photocatalytic H_2 evolution after light-induced Ti^{3+}-O_v states formation. Particularly interesting is that the induction of Ti^{3+}-O_v defects leads to macroscopic observation of powder darkening but the defects presence is detectable only by EPR. Despite the fact that concentration of Ti^{3+}-O_v states must be extremely small, since even XPS spectra do not confirm their presence, those states are able to enhance H_2 evolution without application of any co-catalyst. Finally, application of well shaped particles in a form of nanosheets allow us to verify the lack of amorphous layer formation in the outer part of the particles. Evidently, based on presented results, even minute amount of Ti^{3+} defects, if localized on the surface of the particles, can successfully act as co-catalyst for H_2 evolution.

5. References
[1] Liu N, Steinrück HG, Osvet A, Yang Y, Schmuki P 2017 *Appl. Phys. Lett.* **110** 072102.
[2] Zhou X, Wierzbicka E, Liu N, Schmuki P 2019 *Chem. Commun.* **55** 533–536.
[3] Liu N, Schneider C, Freitag D, Hartmann M, Venkatesan U, Müller J, Spiecker E, Schmuki P 2014 *Nano Lett.* **14** 3309–3313.
[4] Fujishima A, Honda K 1972 *Nature* **238** 37–38.
[5] Fujishima A, Rao TN, Tryk DA 2000 *J. Photochem. Photobiol. C* **1** 1–21.
[6] Schneider J, Matsuoka M, Takeuchi M, Zhang J, Horiuchi Y, Anpo M, Bahnemann DW 2014 *Chem. Rev.* **114** 9919–9986.
[7] De Angelis F, Di Valentin C, Fantacci S, Vittadini A, Selloni A 2014 *Chem. Rev.* **114** 9708–9753.
[8] Linsebigler AM, Guangquan L, Yates JT 1995 *Chem. Rev.* **95** 735–758.
[9] Ni M, Leung MKH, Leung, DYC, Sumathy K 2007 *Renew. Sustain. Energy Rev.* **11** 401–425.
[10] Connelly K, Wahab K, Idriss H 2012 *Mater. Renew. Sustain. Energy* **1** 3.
[11] Zhou X, Liu N, Schmidt J, Kahnt A, Osvet A, Romeis S, Zolnhofer EM, Marthala VRR, Guldi DM, Peukert W, Hartmann M, Meyer K, Schmuki P 2017 *Adv. Mater.* **29** 1604747.
[12] Liu N, Schneider C, Freitag D, Venkatesan U, Marthala VRR, Hartmann M, Winter B, Spiecker E, Osvet A, Zolnhofer EM, Meyer K, Nakajima T, Zhou X, Schmuki P 2014 *Angew. Chem. Int. Ed.* **53** 14201–14205.
[13] Liu N, Zhou X, Nguyen, NT, Peters K, Zoller F, Hwang I, Schneider C, Miehlich ME, Freitag D, Meyer K, Fattakhova-Rohlfing D, Schmuki P 2017 *ChemSusChem* **10** 62–67.
[14] Wierzbicka E, Zhou X, Denisov N, Yoo JE, Fehn D, Liu N, Meyer K, Schmuki P 2019 *ChemSusChem* **12** 1900–1905.
[15] Chen X, Liu L, Yu PY, Mao SS 2011 *Science* **331** 746–50.
[16] Naldoni A, Allieta M, Santangelo S, Marelli M, Fabbri F, Cappelli S, Bianchi CL, Psaro R, Dal Santo V 2012 *J. Am. Chem. Soc.* **134** 7600–7603.
[17] Naldoni A, Altomare M, Zoppellaro G, Liu N, Kment Š, Zbořil R, Schmuki P 2019 *ACS Catal.* **9** 345–364.
[18] Ma Y, Wang X, Jia Y, Chen X, Han H, and Li C 2014 *Chem. Rev.* **114** 9987–10043.

Acknowledgments
The authors would like to acknowledge the Alexander von Humboldt Foundation, ERC, German Research Council (DFG) within the framework of its Excellence Initiative for the Cluster of Excellence ‘‘Engineering of Advanced Materials’’ (www.eam. fau.de) for financial support.

IOP Conf. Series: Materials Science and Engineering **908** (2020) 012002 doi:10.1088/1757-899X/908/1/012002

Study of Zirconia Material Effect on H_2 Production from Ammonia-argon base Gas in Plasma Plate Type Reactor

Mostafa El-Shafie*[1], Shinji Kambara[1], Yukio Hayakawa[1] and Ryoma Sakai[1]

1Gifu University, Environmental and Renewable Energy Systems Division, Graduate School of Engineering, 1-1 Yanagido, Gifu, 501-1193, Japan.
E-mail: mostafaelshafie81@gmail.com

Abstract. In this study, ammonia was decomposed into hydrogen and nitrogen gas by DBD plasma at different ammonia-argon gas base concentrations of 0.5% and 2% in plate type reactor. The effect of zirconia catalytic material on ammonia decomposition was investigated. Furthermore, it was found that the energy density has a positive effect on the outcome of hydrogen gas concentrations for both ammonia concentration experiments. Additionally, it was clear that hydrogen concentration increased with increasing plasma ammonia decomposition residence time. The hydrogen concentration results obtained by plasma-catalyst mode from ammonia concentration of 2% were much higher than that obtained from plasma only experiment. Also, the hydrogen concentration results at different input energy density for NH_3 decomposition with and without the zirconia catalyst system were compared. It can be concluded that the hydrogen concentration of the plasma-catalyst system was much higher than plasma only. It was observed that the addition of zirconia catalytic material into ammonia decomposition process was affected by the physical and electrical properties of the gas discharge. The energy efficiency of both ammonia concentrations experiments was compared, it was seen that the maximum energy efficiencies at NH_3 concentration of 2% was 0.1948 % while at 0.5% NH_3 concentration was 0.084%.

1. Introduction

Hydrogen is often considered as an attractive alternative fuel to overcome the environmental issues. Hydrogen is a clean energy source and can be utilized in portable energy applications with water as the combustion by-product [1]. It can be produced by different renewable energy sources. However, the hydrogen economy is still limited due to the hydrogen storage inability in a safe and economical manner [2]. Hydrogen gas demands are recently growing which motivated the scientific researchers to improve hydrogen production technologies [3-8]. The DBD plasma can create excited and reactive species, which means that high electrons energy are generated allowing them to interact with the catalyst materials at milder temperatures and pressures. The synergistic between catalyst materials and plasma species have been studied to improve the decomposition process of different molecules such as steam reforming of methane, [9] dry reforming of hydrocarbons, [10–14], pollutant decomposition [10, 11], etc.

It has been reported the optical emission spectroscopy (OES) of NH3 decomposition by dielectric barrier discharge plasma (DBD), it has been found that the OES of NH3 was dominated by the band of the electronic excited NH3* molecule, in range of 563.5–567 nm [15–18]. The decomposition of NH_3 gas by DBD plasma could be mainly achieved by the NH_3^* species due to the other species weak band intensities of N_2^*, NH^*, H^*.

$$NH_3 + e \rightarrow NH_3^* + e \quad (1)$$

$$NH_3^* + e \rightarrow NH_2 + H + e \quad (2)$$

$NH_2 + e \rightarrow NH + H + e$ (3)

It was reported that ammonia decomposition by DBD plasma mainly activated to produce NH_3^* species, then activated by an electron to generate NH_2 and NH species. The hydrogen and nitrogen gas can be generated from NH species through the following reaction [16],
$NH + NH \rightarrow N_2 + H_2$ (4)

In this study, we report the hydrogen production measurements from ammonia decomposition of DBD plasma-assisted catalysis on zirconia (ZrO_2) as well as plasma-only reactions to assess hydrogen production through DBD plasma process and distinguish between plasma-only and catalyst material surface contributions. The hydrogen concentrations at different input ammonia-argon base concentrations of 0.5% and 2% are determined. Furthermore, the effect of residence time on hydrogen production is also evaluated. The energy efficiency and output hydrogen flow rates of plasma-assisted zirconia catalytic material are compared.

2. Experimental setup

In this experiment, ammonia-argon base gas was fed at atmospheric pressure into the plate type reactor in a range of 0.1-1.0 L/min and plasma applied voltage ranges of 12-18 kV. The hydrogen production from Ammonia gas concentrations of 0.5% and 2% were monitored. The plasma plate reactor was filled with catalyst material type of zirconia materials, Table (1) shows zirconia material characteristics. While the DBD plasma was ignited in the electrode gap distance of 4.5 mm. Figure (1) shows the schematic diagram of plasma-catalyst assisted for hydrogen production from ammonia gas. Ammonia gas was fed into the plate reactor and was controlled by the mass flow controller. The decomposed gas samples were taken using a syringe and then analyzed by the gas phase chromatography (GC). Also, the hydrogen concentration measurements were carried out for the plate reactor with and without adding zirconia catalytic material to evaluate the catalyst material effect on ammonia decomposition into H_2 and N_2 gas.

Table.1. Characteristics of zirconia catalyst material.

Diameter (mm)	Density (g/mL)	Relative static permittivity (at 10 kHz)	Mass (g)	Volume (L)
1	6.07	20-24	36	0.00593

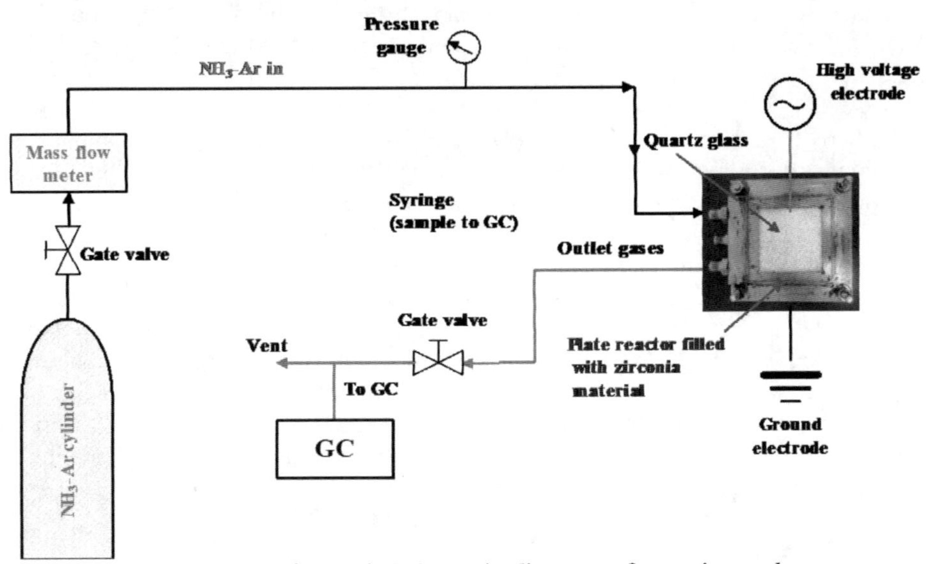

Figure 1. Schematic diagram of experimental setup.

IOP Conf. Series: Materials Science and Engineering **908** (2020) 012002 doi:10.1088/1757-899X/908/1/012002

3. Results & discussions

3.1 *Zirconia effect on H_2 production from 0.5% NH_3-Ar base gas*

Ammonia gas is an important hydrogen energy carrier, additionally, ammonia has no environmental impact because ammonia is carbon-free. Ammonia gas is decomposed by non-thermal plasma or atmospheric to produce hydrogen and nitrogen gas. Zirconia (ZrO_2) catalytic material is used to enhance hydrogen production from ammonia gas. The most interesting observations have been registered where the DBD plasma reactor is filled with the spherical beads of a high dielectric material such as ZrO_2, it can enhance the electric fields at the points of contact between the beads will lead to getting higher concentrations of excited species [19, 20]. Also, it has been reported that the nature of catalyst surface can have an effect on the electric field, and irregularities such as surface roughness or the presence of surface pores can make variations in the electric field and some of the local regions of field intensification can become a high-energy electrons source [21]. The decomposed gases were analyzed by gas phase chromatography (GC), while the outlet gases are collected using a syringe. The conversion rate is determined according to the following expression: *conversion rate [%] = H_2 concentration [%]/ (NH_3 concentration [%] ×1.5) ×100.*
Figure (2) demonstrates the H_2 conversion rate of ammonia gas at different plasma applied voltage and input ammonia flow rates. It was found that the conversion rate increased with plasma voltage at all ammonia input flow rates. Also, the maximum conversion rates of 66.85 %, 33.66%, 26.47% were obtained at ammonia input flow rate of 0.1, 0.5, 1 L/min and plasma voltage of 18 kV, respectively. Furthermore, it was observed that the GC chemical species results included H_2, N_2, O_2 and unreacted ammonia gas. Moreover, there is a positive contribution of the zirconia packing material on the conversion obtained, because of the plasma electric field enhancement at the contact points, but a negative contribution due to the lower residence time of the high ammonia gas flow rates.
Figure (3) shows the hydrogen concentration at different energy density and feeding ammonia flow rates. The energy density (E_d) is calculated according to the following relation: $E_d = ((1000\times P)/v) \times t$, where P is the input power [W], v is the reactor volume [cm^3], and ammonia decomposition residence time [s]. It was found that the hydrogen concentration increased with the energy density increased. Also, it was observed that the maximum obtained hydrogen concentration was 0.499% at ammonia input flow rate of 0.1 L/min and energy density of 69.6 J/cm^3. It was clear that the best results were obtained at low NH_3 input flow rate due to high energy density. Figure (4) shows the residence time effect on hydrogen production from ammonia decomposition at DBD plasma applied voltage of 18 kV. It was clear that residence time was an important factor for ammonia decomposition into hydrogen and nitrogen gas which appeared with increasing ammonia input flow rate. It can be concluded that residence time has a negative effect on the hydrogen production from ammonia decomposition by DBD plasma influencing the retention time which changes the electric field distribution and plasma reactivity.

3.2 *Zirconia effect on H_2 production from 2% NH_3-Ar base gas*

To understand ammonia decomposition using plasma-catalysts, it is important to ammonia decomposition reaction mechanism which is initiated by the adsorption of ammonia molecules onto the active site surface. These adsorbed ammonia molecules undergo successive N–H bond cleavage, releasing hydrogen atoms that can combine to form molecular hydrogen [22]. Neyts and Bal [23] have recently investigated that the catalytic material addition to a plasma discharge can improve the plasma electric field influencing the species retention time at the catalyst material surface and increasing the surface reactivity. Moreover, the catalyst materials will yield more electrical energy when materials of higher dielectric constant are used resulting in the formation of more electrons of higher energy which increased the ionized and excited species. In this work, the effect of the ammonia feeding gas concentration on hydrogen production was investigated. Also, zirconia catalytic material was utilized to fill the plasma plate reactor type. Ammonia —argon base gas with a concentration of 2% was fed into the reactor at 0.1 L/min and atmospheric plasma voltage range of 12—18 kV.

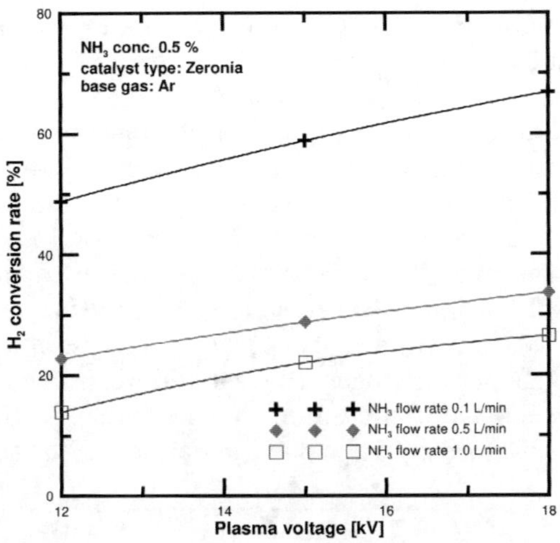

Figure 2. The effect of plasma voltage on H_2 conversion at different NH_3 flow rates.

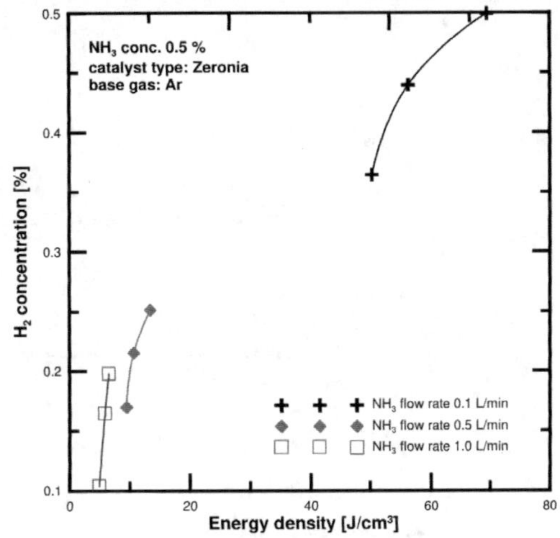

Figure 3. Energy density effect on H_2 concentration at different NH_3 flow rates.

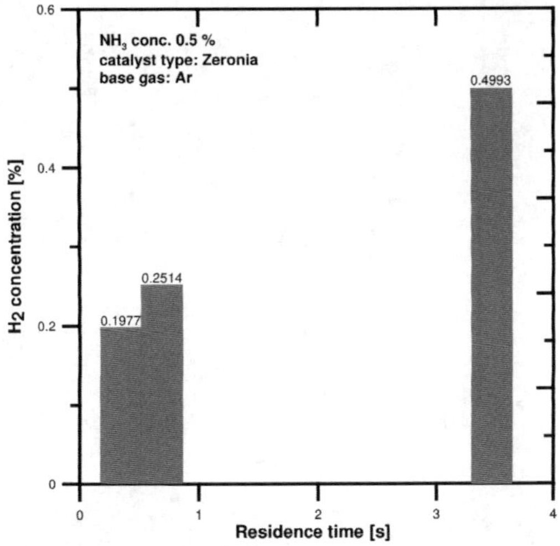

Figure 4. Residence time effect on hydrogen concentration at plasma voltage of 18 kV.

IOP Conf. Series: Materials Science and Engineering **908** (2020) 012002 doi:10.1088/1757-899X/908/1/012002

Also, this experiment is carried out without catalyst materials to declare zirconia catalytic material effect on the hydrogen production from ammonia-Ar gas base at different plasma applied voltage. Ammonia-Ar base experiment was carried out with and without filling the plate reactor with zirconia catalytic material. Figure (5) illustrates the hydrogen conversion rate results of both experiments with and without adding zirconia catalytic material. It was found that H_2 conversion rate increased with the plasma applied voltage increased. Also, the conversion rate results of ammonia decomposition experiment with adding zirconia catalytic material have higher values than ammonia decomposition experiment without catalyst material. It can be concluded that zirconia material can be utilized to enhance ammonia decomposition using DBD plasma. To clarify the catalytic material effect on H_2 production from ammonia gas, the hydrogen concentration versus the energy density is shown in figure (6). It was observed that the hydrogen concentration enhanced by the addition of zirconia catalytic material to the ammonia decomposition system by DBD plasma. It was clear that the ammonia gas molecules are absorbed by zirconia material active surface producing higher electron energy.

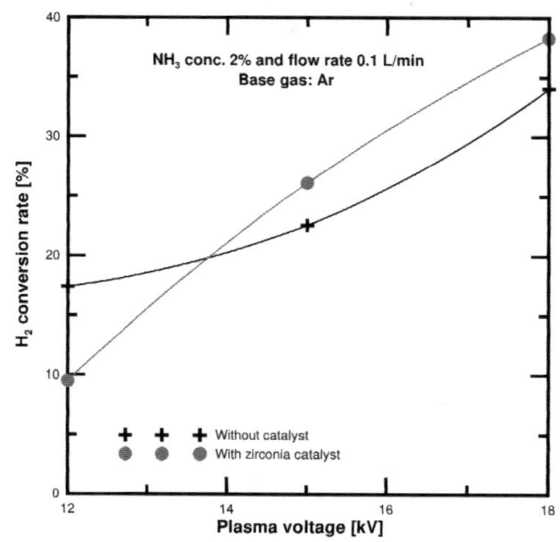

Figure 5. Comparison of zirconia catalyst effect on H_2 conversion rate from 2% NH_3.

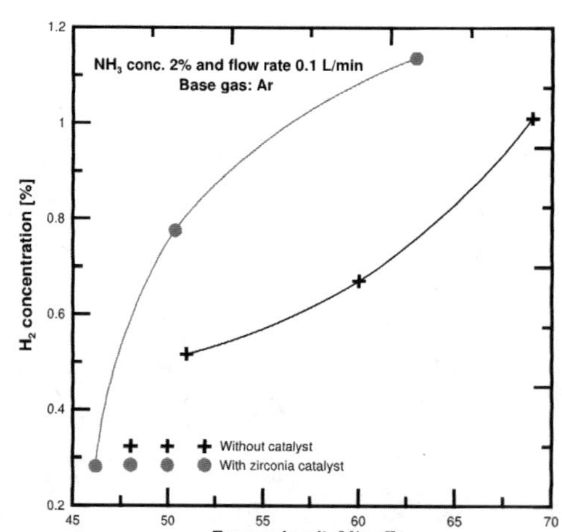

Figure 6. Comparison of H_2 concentrations with and without zirconia catalyst at different energy density.

3.3 *Evaluation of ammonia-Ar base gas decomposition by DBD plasma*

To assess the decomposition process of the ammonia-Ar gas base at different concentrations by the DBD plasma, the hydrogen flow rate was determined at an input flow rate of 0.1 L/min and at different input electric power. A comparison between the hydrogen flow rates results of ammonia input concentrations of 0.5 & 2% was described in figure (7). It was found that the hydrogen flow rate increased with the input power for both ammonia input concentrations. However, the hydrogen flow rate results of the ammonia input concentration of 2% were higher than lower input ammonia concentration at different input electric power. In addition to the catalyst effect on ammonia decomposition by DBD plasma, it can be concluded that the hydrogen production will also depend on the plasma input power. The energy efficiency of the DBD ammonia decomposition process will also depend on the catalyst material structural features such as its morphology, porosity, and chemical activity as well as plasma nature. Figure (8) compares the energy efficiency results of both ammonia experiment concentrations at different electric power. It was clear that energy efficiency increased with increasing plasma input power. Furthermore, it was found that the maximum energy efficiency obtained at NH_3 concentration of 2% was 0.1948 % while at 0.5% NH_3 concentration was 0,084%.

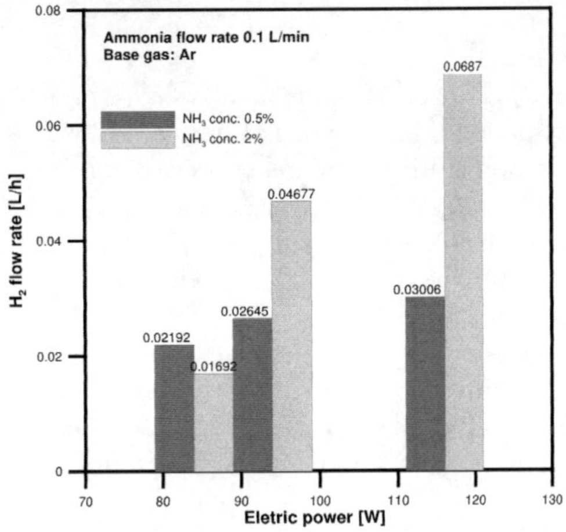

Figure 7. Comparison between H_2 flow rate results of different ammonia input concentrations.

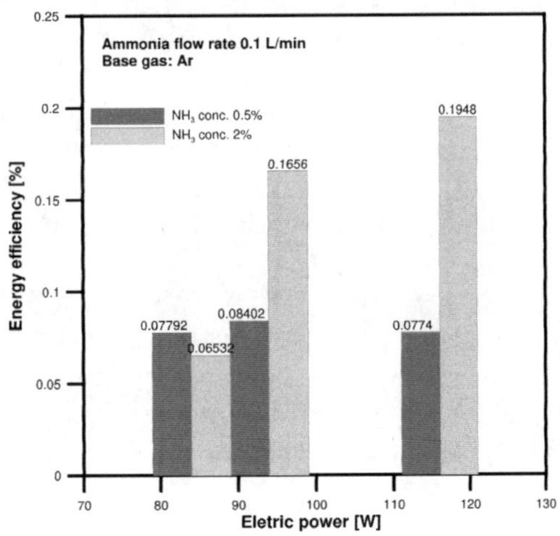

Figure 8. Comparison of energy efficiency results at different plasma input power.

4. Conclusion

Ammonia-argon base gas was decomposed into hydrogen and nitrogen gas by DBD plasma at different ammonia-argon gas base concentrations of 0.5% and 2% were investigated. The effect of zirconia catalyst material on ammonia decomposition was studied. It was found that the NH_3 conversion rate was significantly increased when DBD plasma was combined with the catalyst. Furthermore, it was observed that the energy density has a positive effect on the produced hydrogen gas concentrations of both ammonia concentrations experiments. Additionally, it was clear that

IOP Conf. Series: Materials Science and Engineering **908** (2020) 012002 doi:10.1088/1757-899X/908/1/012002

hydrogen concentration increased with ammonia decomposition reaction time or residence time increase. The NH_3 conversion and hydrogen concentration obtained by plasma-catalyst mode from ammonia concentration of 2% were much higher than those obtained using plasma only. Also, it was compared the hydrogen concentration results at different input energy density for NH_3 decomposition with and without zirconia catalyst. It was found that the hydrogen concentration of plasma-catalyst system was much higher than plasma only system. An evaluation of plasma-catalyst system was developed to compare different ammonia input concentrations. In general, adding zirconia catalytic material into the ammonia decomposition process by DBD plasma will affect the physical and electrical properties of the gas discharge which causes some modification to the produced hydrogen and nitrogen gases of the plasma-catalytic processes. The energy efficiency results of both ammonia concentrations experiments were compared. It was seen that the maximum energy efficiency obtained at NH_3 concentration of 2% was 0.1948 % while at 0.5% NH_3 concentration was 0.084%.

5. References

[1] Yin SF, Xu BQ, Zhou XP, Au CT (2004) Appl Catal A Gen 277:1–9.
[2] Hill A, Torrente L (2014) Murciano. Int J Hydrog Energy 39:7646–7654.
[3] Rosen, M. A.; Scott, D. S. Comparative efficiency assessment for a range of hydrogen production processes. *Int. J. Hydrogen Energy* 1998, *23*, 631-640.
[4] L. Barelli, G.Bidini, F.Gallorini, S.Servili, Hydrogen production through sorption-enhanced steam methane reforming and membrane technology: a review, Energy 33(2008)554–570.
[5] Shiga, H.; Shinda, K.; Hagiwara, K.; Tsutsumi, A.; Sakurai, M.; Yoshida, K.; Balgen, E. Large scale hydrogen production from biogas. *Int. J. Hydrogen Energy* **1998**, *23*, 631-640.
[6] J. D. Holladay, J. Hu, D. L. King, Y. Wang, An overview of hydrogen production technologies, Catal. Today 139 (2009) 244-260.
[7] M. El-Shafie, S. Kambara, Y. Hayakawa, T. Miura. Preliminary results of hydrogen production from water vapor decomposition using DBD plasma in a PMCR reactor. Int. J. Hydrogen Energy, 2019; 44: 20239-48.
[8] M. El-Shafie, S. Kambara, Y. Hayakawa, Hydrogen Production Technologies Overview. Journal of Power and Energy Engineering 7(2019) 107-154.
[9] Nozaki, T.; Okazaki, K. Non-thermal plasma catalysis of methane: Principles, energy efficiency, and applications. Catal. Today 2013, 211, 29−38.
[10] Allah, Z. A.; Whitehead, J. C. Plasma-catalytic dry reforming of methane in an atmospheric pressure AC gliding arc discharge. Catal. Today 2015, 256, 76−79.
[11] Gallon, H. J.; Tu, X.; Whitehead, J. C. Effects of Reactor Packing Materials on H 2 Production by CO 2 Reforming of CH 4 in a Dielectric Barrier Discharge. Plasma Processes Polym. 2012, 9, 90−97.
[12] Tu, X.; Gallon, H. J.; Twigg, M. V.; Gorry, P. A.; Whitehead, J. C. Dry reforming of methane over a Ni/Al2O3 catalyst in a coaxial dielectric barrier discharge reactor. J. Phys. D: Appl. Phys. 2011, 44 (27), 274007−274007.
[13] Tu, X.; Whitehead, J. C. Plasma-catalytic dry reforming of methane in an atmospheric dielectric barrier discharge: Understanding the synergistic effect at low temperature. Appl. Catal., B 2012, 125, 439−448.
[14] Kim, J.; Abbott, M. S.; Go, D. B.; Hicks, J. C. Enhancing C−H Bond Activation of Methane via Temperature-Controlled, Catalyst− Plasma Interactions. ACS Energy Lett. 2016, 1 (4), 94−99.
[15] Watson, J. K. G., & Majewski, W. A. (1986). Assignment of the schuster band of ammonia. Journal of Molecular Spectroscopy, 115, 82–87.
[16] Agostino, R., Cramarossa, F., Benedictis, S. D., & Ferraro, G. (1981). Kinetic and spectroscopic analysis of NH3 decomposition under R.F. plasma at moderate pressure. Plasma Chemistry and Plasma Processing, 1, 19–35.
[17] Cicala, G., De Tommaso, E., Raino, A. C., Lebedev, Y. A., & Shakhatov, V. A. (2009). Study of positive column of glow discharge in nitrogen by optical emission spectroscopy and numerical simulation. Plasma Sources Science and Technology, 18, 025032.

[18] Nicholas, J. E., Spiers, A. I., & Martin, N. A. (1986). Kinetics and mechanism in the decomposition of NH3 in a radio-frequency pulse discharge. Plasma Chemistry and Plasma Processing, 6, 39–51.

[19] Zhang, Y., Wang, H., Jiang, W., & Bogaerts, A. (2015). Two-dimensional particle-in cell/ Monte Carlo simulations of a packed-bed dielectric barrier discharge in air at atmospheric pressure. New Journal of Physics, 17, 083056.

[20] Koen Van, L., & Annemie, B. (2016). Fluid modelling of a packed bed dielectric barrier discharge plasma reactor. Plasma Sources Science and Technology, 25, 015002.

[21] Butterworth, T., Elder, R., & Allen, R. (2016). Effects of particle size on CO2 reduction and discharge characteristics in a packed bed plasma reactor. Chemical Engineering Journal, 293, 55–67.

[22] Hansgen DA, Vlachos DG, Chen JG (2010) Nat Chem 2:484–489.

[23] Neyts, E. C., & Bal, K. M. (2017). Effect of electric fields on plasma catalytic hydrocarbon oxidation from atomistic simulations. Plasma Processes and Polymers, 14, 1600158.

Acknowledgments
The authors would like to thank ***Prof. Shinji Kambara*** (Environmental and energy systems department-Gifu University-Japan) for his continuous support.

IOP Conf. Series: Materials Science and Engineering **908** (2020) 012003 doi:10.1088/1757-899X/908/1/012003

TiO_2 Synthesis Anatase Phase with The Sol-gel Process at Room Temperature

P Argüelles-Lucho[1], R Pérez-Cuapio[2], R M Woo-Garcia[1], L García-González[1], M. Pacio[3] and F López-Huerta[4,5]*

1 Centro de Investigación en Micro y Nanotecnología, Universidad Veracruzana, Veracruz, 94294, Boca del Río, Veracruz, México.
2 Benemérita Universidad Autónoma de Puebla, Facultad de Ingeniería Química, Av. Sn. Claudio y Blvd. 18 sur, Jardines de San Manuel, 72570, Puebla, México.
3 Instituto de Ciencias CIDS, Benemérita Universidad Autónoma de Puebla, 14 Sur y Av. San Claudio, C.U., Edif. IC-5, San Manuel 72570, Puebla, México.
4 Facultad de Ingeniería Eléctrica y Electrónica, Universidad Veracruzana, Veracruz, Boca del Río, Veracruz, México.
5 Maestría en Ingeniería Aplicada, Facultad de Ingeniería de la Construcción y el Hábitat, Universidad Veracruzana, Calzada Ruíz Cortines 455, Boca del Río, Veracruz 94294, México.
E-mail: frlopez@uv.mx

Abstract. Titanium oxide (TiO_2) film was deposited on 304 stainless steel substrates using the spin coating technique. The synthesis of the deposited solution was with the sol-gel method at room temperature, using titanium isopropoxide as a precursor. Obtained film was thermally treated at temperatures of 500 °C, 550 °C and 600 °C. The X-ray diffraction (XRD) analysis confirmed the crystallinity of the film in the anatase phase, with the highest intensity at the plane (101). Moreover, the crystallite size increased with the rising temperature of the thermal treatment. The thickness of the deposited film was obtained with profilometry being: 1 μm for the 500 °C film, 1.2 μm for the 550 °C film, and 1.5 μm for the 600 °C film. Raman spectroscopy measurements showed vibrational modes at 144, 399, 519 and 638 cm^{-1}, characteristic of the TiO_2 anatase phase. The obtained results showed that these films have a high potential for gas sensor applications due to their surface-to-volume ratio after obtaining the size of nanometric crystallite.

1. Introduction

Respiratory illnesses in the last few years have increased due to volatile organic compound (VOC) emissions [1]. Certain VOCs have pleasant odors or are odorless, making them difficult to detect. They can cause poisoning and cancer, therefore a gas sensor which can detect human physical integrity is necessary. There are different gas sensors which are covered with titanium oxide (TiO_2) nanostructured film with its oxygen vacancy mechanism (redox), as well as its ionization mechanism [2-4]. TiO_2 is a type n semiconductor with a bandwidth prohibited from 3 to 3.6 eV. TiO_2 is not toxic, can be used in hostile environments, and has three crystalline phases: anatase, rutile (both tetragonal), and brookite (orthorhombic). The anatase crystalline structure is the most commonly used for gas sensors because of its greater reactivity and simple crystalizing structure, starting at 400 °C [5]. TiO_2 can be obtained with different methods, the most utilized being chemical vapor deposition (CVD), precipitation, sputtering, and sol-gel [4, 5, 7]. The sol-gel method is a chemical synthesis that offers great benefits compared to other techniques. It has a low cost, good homogeneity, film thickness control, room temperature deposits,

IOP Conf. Series: Materials Science and Engineering **908** (2020) 012003 doi:10.1088/1757-899X/908/1/012003

and porous film. The most commonly used precursor for film formation is a metallic alkoxide dissolved in hydrolyzed alcohol in an acidic, neutral, or basic environment [7,8]. TiO_2 film properties deposited with spin coating can vary, depending on centrifuge velocity, time, and the evaporation of the solution. Moreover, it can be deposited on different substrates such as glass, silicon, and 304 stainless steel [7,9,10]. 304 stainless steel was selected for this study due to its good properties for the implementation of oscillate structures. 304 stainless steel can withstand thermal treatments of up to 900 °C, has an accessible price, and mechanical properties similar to silicon, which is why it is used as a substrate for gas sensors. The synthesis and the results of multilayer TiO_2 film deposited in 304 stainless steel substrates with the spin coating method annealed at different temperatures are presented in the following sections.

2. Methodology

TiO_2 was synthesized using titanium isopropoxide (TTIP) as a precursor with an ethanol concentration of 0.0025 M. Hydrocloric acid (HCl) was added at a concentration of 0.0014 M. This mix was maintained in continuous agitation at room temperature for 10 minutes until a transparent and homogeneous solution was obtained. The obtained solution was deposited on 304 stainless steel substrates in an area of 2 cm by 2 cm with spin coating at 2000 revolutions per minute (rpm) for 17 seconds at room temperature. The deposited film dried at 200 °C for 5 minutes to eliminate organic solvent, and this deposit-drying process was repeated 10 times.

Finally, the dry film was annealed at 500 °C, 550 °C, and 600 °C for 2 hours. The obtained film was analyzed by means of X-ray diffraction with the Bruker D8 Advance, with CuKα radiation. The vibrational modes of TiO_2 film were obtained with Raman spectroscopy using a 623.8 nm Neon (He-Ne) helium laser. The profilometry measurements were taken with a Veeco Dettak 150 profilometer, which scanned with a micro-point 3 mm above the surface.

3. Results and discussion

The TiO_2 film deposited on the 304 stainless steel substrates with thermal treatments of 500 °C, 550 °C, and 600 °C had excellent uniformity and qualitative adherence. Figure 1 shows the X-ray diffraction results for the TiO_2 film on 304 stainless steel substrates with a variation of thermal treatments at 500 °C, 550 °C, and 600 °C, showing tetragonal anatase phase, which conforms to PDF #01-072-7058. All the diffractograms show the film have the most intensity at the plane (101). As the temperature at the crystalline plane increases, a slight shift in tension can be seen.

Crystal size was theoretically calculated with the Scherrer equation by using the peak (101) [11]:

$$L = \frac{k\lambda}{\beta \cos \theta} \tag{1}$$

The crystallite size for the 500 °C TiO_2 sample is 24.7 nm, the 550 °C TiO_2 sample is 25.4 nm, and the 600 °C TiO_2 sample is 26.6 nm. The nanometric crystallite size allows us to apply this film to objects such as gas sensors, due to the strong interaction between the TiO_2 nanoparticles and the gas to be detected and the surface-to-volume ratio between them. Film thickness was obtained with profilometry measurements, giving us the following results: 1 μm, 1.2 μm and 1.5 μm for the annealed film at temperatures of 500 °C, 550 °C and 600 °C respectively.

The tension shifts were analyzed with the Stoney equation [12] for TiO_2 film with thermic treatments of 550 °C and 600 °C at 2 theta (2θ), the shifts were found at 25.20 and 25.19 °.

$$\sigma_s = -\frac{E}{\mu}\frac{d - d_0}{d_0} \tag{2}$$

Location: E is the Young module (151 Gpa), μ is the TiO_2 Poisson coefficient (0.27) [13], d_0 is the tension-free network reference (3.518310 Å), conforming to the PDF #01-072-7058 document, and d is the interplanar sample distance to be analyzed. The results were 2.26 GPa for 550 °C and 2.04 GPa for 600 °C.

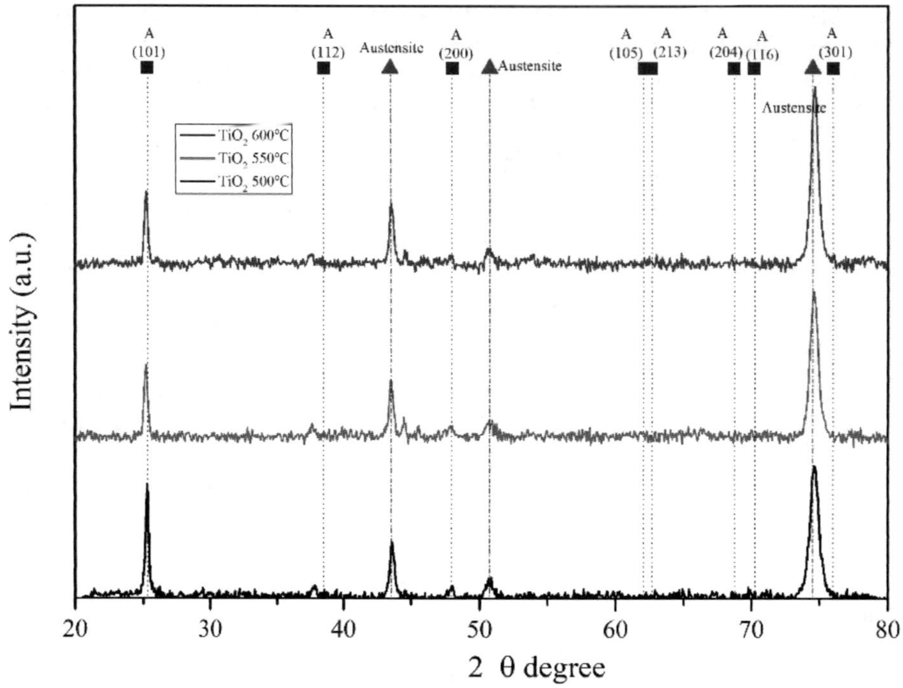

Figure 1. X-ray diffractograms of TiO_2 anatase phase film with thermal treatments of 500 °C, 550 °C and 600 °C.

Figure 2 shows Raman spectrums for the deposited film treated at different temperatures. These spectrums have vibrational modes of E_g (144 cm^{-1} and 638 cm^{-1}), B_{1g} (399 cm^{-1} and 519 cm^{-1}), and A_{1g} (519 cm^{-1}), confirming the presence of TiO_2 anatase phase [10]. This reveals a compression shift and better crystallinity at vibrational mode E_g (144 cm^{-1}), which is associated with a particle size of less than 40 nm, matching the grain size calculated from the TiO_2 film [14].

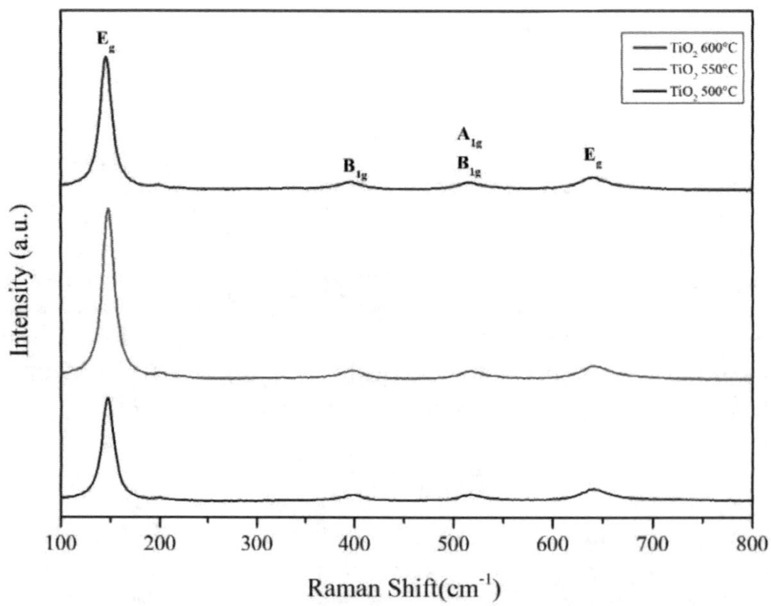

Figure 2. Raman spectrum of TiO_2 anatase phase multilayer film with thermal treatments of 500 °C, 550 °C and 600 °C.

IOP Conf. Series: Materials Science and Engineering **908** (2020) 012003 doi:10.1088/1757-899X/908/1/012003

4. Conclusions

A TiO_2 solution was synthesized using the sol-gel route at room temperature. The TiO_2 solution was deposited on 304 stainless steel substrates with high homogeneity and good adherence on the substrate. The obtained X-ray diffraction and Raman spectroscopy results confirm the presence of TiO_2 at the anatase phase with a preferential growth at the plane (101). The crystallinity of the TiO_2 film varied in function with the temperature. The deposited film in this study has a high potential to be used for gas sensors due to the preferential growth at the plane (101) and the presence of the tetragonal anatase phase. The nanometric crystallite size allows for better interaction between the TiO_2 film and the gas to be detected with there is a greater contact surface area.

As future work, studies of characterization as photoluminescence, nanoindentation, and performance of the multilayer TiO_2 films will be realized in a gas chamber with different VOCs.

5. References

[1] World Health Organization (WHO). An estimated 12.6 million deaths each year are attributable to unhealthy environments. Available online: https://www.who.int/news-room/detail/15-03-2016-an-estimated-12-6-million-deaths-each-year-are-attributable-to-unhealthy-environments (accessed on 12 August 2019).

[2] Sarfraz, S., & Kumar, R. V. (2014, January). Ultrasonically assisted synthesis of mesoporous TiO 2 nanoparticle thin films via sol-gel process for chemo-resistive gas sensor applications. In *Proceedings of 2014 11th International Bhurban Conference on Applied Sciences & Technology (IBCAST) Islamabad, Pakistan, 14th-18th January, 2014* (pp. 48-51). IEEE.

[3] Gurlo, A., & Riedel, R. (2007). In situ and operando spectroscopy for assessing mechanisms of gas sensing. *Angewandte Chemie International Edition*, *46*(21), 3826-3848.

[4] MiarAlipour, S., Friedmann, D., Scott, J., & Amal, R. (2018). TiO2/porous adsorbents: Recent advances and novel applications. *Journal of hazardous materials*, *341*, 404-423.

[5] Li, Z., Yao, Z., Haidry, A. A., Plecenik, T., Xie, L., Sun, L., & Fatima, Q. (2018). Resistive-type hydrogen gas sensor based on TiO2: A review. *International Journal of Hydrogen Energy*, *43*(45), 21114-21132.

[6] Löbl, P., Huppertz, M., & Mergel, D. (1994). Nucleation and growth in TiO2 films prepared by sputtering and evaporation. *Thin solid films*, *251*(1), 72-79.

[7] Brinker, C. J., Hurd, A. J., Schunk, P. R., Frye, G. C., & Ashley, C. S. (1992). Review of sol-gel thin film formation. *Journal of Non-Crystalline Solids*, *147*, 424-436.

[8] Sayilkan, F., ASİLTÜRK, M., Sayilkan, H., Önal, Y., Akarsu, M., & Arpaç, E. (2006). Characterization of TiO_2 Synthesized in Alcohol by a Sol-Gel Process: The Effects of Annealing Temperature and Acid Catalyst. *Turkish Journal of Chemistry*, *29*(6), 697-706.

[9] Bakri, A. S., Sahdan, M. Z., Adriyanto, F., Raship, N. A., Said, N. D. M., Abdullah, S. A., & Rahim, M. S. (2017, January). Effect of annealing temperature of titanium dioxide thin films on structural and electrical properties. In AIP conference proceedings (Vol. 1788, No. 1, p. 030030). AIP Publishing.

[10] Rattana, T., Witit-Anun, N., Suwanboon, S., & Chaiyakun, S. (2014). Preparation and properties of TiO2 thin films deposited on different substrates by sol-gel method. In *Advanced Materials Research* (Vol. 979, pp. 355-358). Trans Tech Publications.

[11] Kim, D. J., Hahn, S. H., Oh, S. H., & Kim, E. J. (2002). Influence of calcination temperature on structural and optical properties of TiO2 thin films prepared by sol–gel dip coating. Materials Letters, 57(2), 355-360.

[12] Espinoza-Beltrán, F. J., Che-Soberanis, O., Garcıa-González, L., & Morales-Hernández, J. (2003). Effect of the substrate bias potential on crystalline grain size, intrinsic stress and hardness of vacuum arc evaporated TiN/c-Si coatings. Thin Solid Films, 437(1-2), 170-175.

[13] Borgese, L., Gelfi, M., Bontempi, E., Goudeau, P., Geandier, G., Thiaudière, D., & Depero, L. E. (2012). Young modulus and Poisson ratio measurements of TiO2 thin films deposited with Atomic Layer Deposition. Surface and Coatings Technology, 206(8-9), 2459-2463.

IOP Conf. Series: Materials Science and Engineering **908** (2020) 012003 doi:10.1088/1757-899X/908/1/012003

[14] Li Bassi, A., Cattaneo, D., Russo, V., Bottani, C. E., Barborini, E., Mazza, T., ... & Pratsinis, S. E. (2005). Raman spectroscopy characterization of titania nanoparticles produced by flame pyrolysis: the influence of size and stoichiometry. Journal of Applied Physics, 98(7), 074305.

Acknowledgments

This research was funded by project PRODEP "Modelado y Desarrollo de Microsensores CMOS-MEMS para la detección de diabetes" and CONACYT trough Grant 517529 by Program of Doctorate in Materials and Nanoscience of the Research Center in Micro and Nanotechnology of University of Veracruz (MICRONA-UV), México.

IOP Conf. Series: Materials Science and Engineering **908** (2020) 012004 doi:10.1088/1757-899X/908/1/012004

Shrinkage and Swelling of Greek Chestnut Wood (*CastaneaSativaMill.*) in Relation to Extractives Presence

M Chavenetidou[1], K V Kakavas[2] and D Birbilis[3]

1Laboratory of Forest Utilization, School of Forestry and Natural Environment, Aristotle University of Thessaloniki, P.O. Box 247, 54124 Thessaloniki, Greece, marinasmith13@gmail.com
2Laboratory of analytical and environmental chemistry, Department of Environment, University of Thessaly, Geopolis – Larisa, Greece, orcid 0000-0001-8247-0134, email: kakavas@uth.gr
3Department of Forestry, Wood Science and Design, University of Thessaly, Karditsa, Greece, dbirbilis77@gmail.com
Email: marinasmith13@gmail.com

Abstract. The study of various woods shrinkage is of particular practical importance because the wood's property to shrink and swell with the adsorption or absorption of moisture from the atmosphere is a main cause of many defects that occur in wood and furniture during weather conditions changing. In the present study, measurements of the shrinkage of chestnut coppice wood samples were performed before and after extraction. The extraction was carried out with hot water for 6 and 12 days. Afterwards, the dry density, radial and tangential shrinkage were determined both in air-dry conditions and in absolute dry conditions. Moreover, the fiber saturation point was determined for the selected samples. The most significant result arrived from this study is that, now, in conjunction with the other researches have been made, there is enough evidence for chestnut wood from Mediterranean coppice forests, to claim that maximum radial shrinkage is about 3.83% (ranging from 3.40% - 4.12%) and maximum tangential shrinkage is about 6.58% (ranging from 6.21% - 8.20%).

1. Introduction

Chestnut (*Castanea sativa Mill*) is a precious forestry species of Greece, that occupies an area of 0.51% from the total Greek forest area and produces 8.58% of the total technical hardwood. In coppice forest (cutting age around 20-25 years) produces timber that is mainly used as construction wood and furniture. Shrinkage and swelling of wood are important properties, which affect its utilization. The effect might be a simple fluctuation of dimensions or an uneven shrinkage or swelling due to different humidity or density distribution. These results in various defects such as alteration of wooden objects shape, warping, opening or tightening of joints, change of cross-sectional shape, checking or even collapse of wood [1]. Among the factors influencing shrinkage (density, humidity, extractives), the influence of extractives [2] was studied in particular. Tangential wood shrinkage measurements, according to other studies [3-5] ranged from 2,8 to 10%, while in our study it ranged from 4 to 6,5%.

2. Materials and methods

Shrinkage was calculated in green wood specimens which came from logs cut from the base of five (5) chestnut coppice trees aged 25-27 years harvested in Arnaia (region of Halkidiki, Greece). Each log was cut to boards of 3 cm thick approximately. From every log one of the two central boards were selected and used for the creation of shrinkage specimens. From each board (a, b, c, d, e and f) the shrinkage specimens were formed with successive tangential cuts from pith to bark and in the distance of 2-4 cm between them. Figure 1 shows the way of specimens' formation from each log, their position related to pith, the direction of the annual rings and codes as well. In total, 198 samples were formed (Log A and B gave 36, Log C, D and E gave 42). The samples dimensions for the determination of shrinkage (radial, tangential) were measured successively from initial green condition, after air drying and then to absolute dry condition (kiln drying to 103 °C), with the aid of electronic micrometre. For the determination of humidity content of air-dried specimens their weight and their dry dimensions were measured at the same time.

Logs A, B

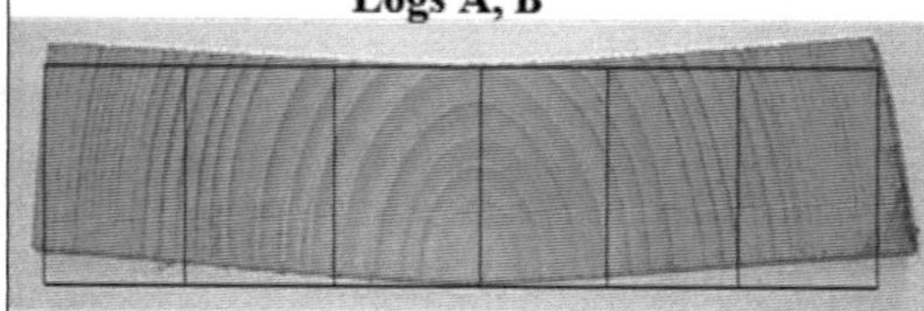

a	21	20	19	1	2	3
b	24	23	22	4	5	6
c	27	26	25	7	8	9
d	30	29	28	10	11	12
e	33	32	31	13	14	15
f	36	35	34	16	17	18

Log C

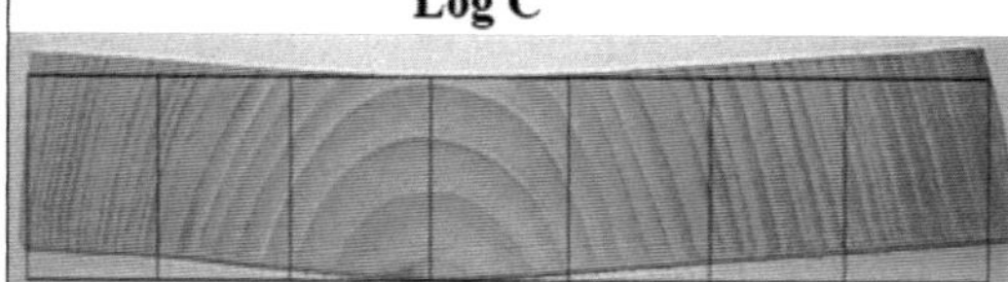

a	27	26	25	1	2	3	4
b	30	29	28	5	6	7	8
c	33	32	31	9	10	11	12
d	36	35	34	13	14	15	16
e	39	38	37	17	18	19	20
f	42	41	40	21	22	23	24

Logs D, E

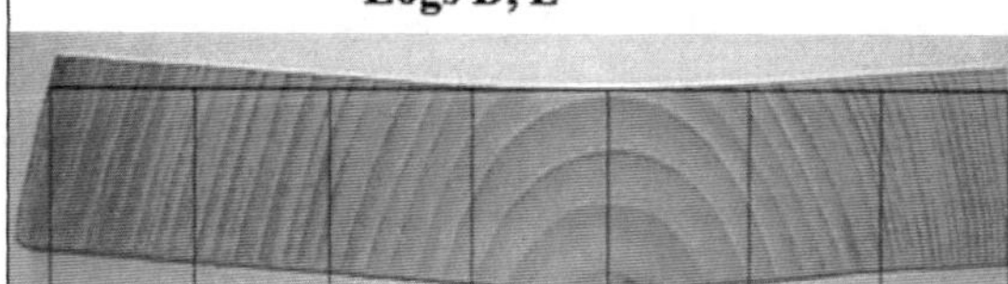

a	22	21	20	19	1	2	3
b	26	25	24	23	4	5	6
c	30	29	28	27	7	8	9
d	34	33	32	31	10	11	12
e	38	37	36	35	13	14	15
f	42	41	40	39	16	17	18

Figure 1. Way of samples formation from each log and their codes (air dried)

The specimens volume used for the determination of dry density was calculated by their geometrical dimensions. The extractives effect was studied to specimens coded with a1, a2, a3, c7, c8, c9, b4, b5, b6 and d10, d11, d12, after the determination of shrinkage. Samples a and c were extracted with hot water for 6 days, when b and d for 12 days. The extractions were conducted to a Soxhlet device for 4 hours daily. The rest of the time wood samples were sunk into water.

3. Results and discussion

From the radial and tangential dimensions of the specimens (in wet, air-dry and at zero moisture content conditions), the shrinkage was determined based on the ratio (1):

$$\beta(\%) = (l_1 + l_2) \times 100/l_1 \quad (1)$$

l_1, initial green dimension (cm)
l_2, final dry dimension (cm)

Swelling [1] can be determined by the relation (2):

$$\alpha(\%) = 100 \times \beta/1 - \beta \quad (2)$$

α, swelling (%)

β, shrinkage (%)

Considering that the relationship between the changes in wood dimensions and the moisture content up to the fiber saturation point is rectilinear, the fiber saturation point was determined by the changes in the radial and tangential dimensions of the specimens (from the wet conditions to the air-dry and zero moisture content conditions) on the basis of the relation (3):

$$\beta_1/\beta_2 = f/f - y \quad (3)$$

β_1, maximum shrinkage (%)
β_2, air-dry shrinkage (%)
f, fiber saturation point (%)
y, air-dry equivalent moisture content (%).

The determination of the samples dimensions was held with the use of a micrometer and repeated measurements until the stabilization of the dimensions. From each specimen seven smaller parts came out and were measured, after they were fully saturated.Dry density, radial and tangential shrinkage and fiber saturation point are shown in Table 1.As shown in Table 1, the mean maximum radial shrinkage of chestnut wood was 3.83% and ranged from 3.58% - 4.12%. Tangential shrinkage was 6.58% and ranged from 6.21% -7.01%. The shrinkage values determined were comparatively similar to corresponding values in the literature [6-8], with the exception of the large tangential shrinkage value (10.0%) reported by Fioravanti [9]. There were no differences between the shrinkage values of the present study and the values from the literature. As shown in Table 1, the fiber saturation point calculated from changes in the tangential dimensions of the samples and their moisture content was found to be slightly higher in comparison to corresponding values determined by the changes in radial dimensions. This difference led to first changing the tangential dimensions of the wood and then the radial ones when reducing moisture content. This difference is very low and does not affect the commercial use of wood.

Table 1. Dry density, radial/tangential shrinkage and fiber saturation point (FSP) of chestnut wood (Logs A, B, C, D and E).

Sample	Dry density (g/cm^3)	Shrinkage (%)*		Shrinkage (%)**		FSP (%) ***	
		Rad.	Tang.	Rad.	Tang.	Rad.	Tang.
$A_{1\text{-}36}$	0,577	2,399	3,821	4,122	6,374	26,574	27,560
$B_{1\text{-}36}$	0,551	2,095	3,637	3,754	6,209	24,680	26,007
$C_{1\text{-}42}$	0,544	2,204	4,406	3,610	7,013	29,008	30,051
$D_{1\text{-}42}$	0,619	2,069	4,237	3,582	6,938	26,985	29,196
$E_{1\text{-}42}$	0,575	2,376	3,869	4,075	6,357	27,250	29,288
Avg.	0,573	2,229	3,994	3,829	6,578	26,899	28,420

* Shrinkage from wet (green) to air-dry condition (moisture content 10,615% - 11,640%)
** Maximum Shrinkage from wet (green) to oven-dry condition (moisture content 0%)
*** Determination from dimensions changes (radial and tangential respectively) according to FSP (fiber saturation point)

The radial variation (from pith to bark) of the maximum shrinkage for each log is shown separately in Figure 2. As shown in Fig. 2 the radial shrinkage decreases from pith to bark (logs C, D and E). In the logs A and B, shrinkage seemed to either slight decrease or initially to increase and then decrease along with the radial direction from pith to bark. Tangential shrinkage was found to increase in the direction from pith to bark (logs A and E) or to decrease (logs B, C and D).

IOP Conf. Series: Materials Science and Engineering **908** (2020) 012004 doi:10.1088/1757-899X/908/1/012004

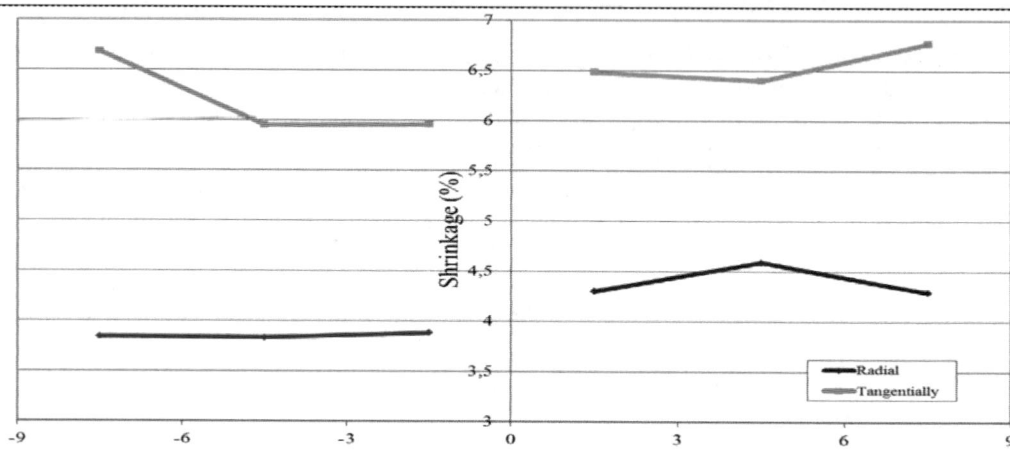

Figure 2a. Block A, distance from pith (cm). Upper tangentially and down radial

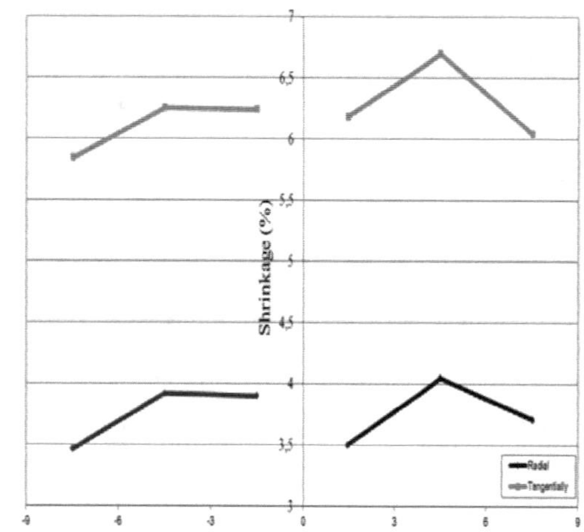

Figure 2b. Block B, distance from pith (cm). Upper tangentially and down radial

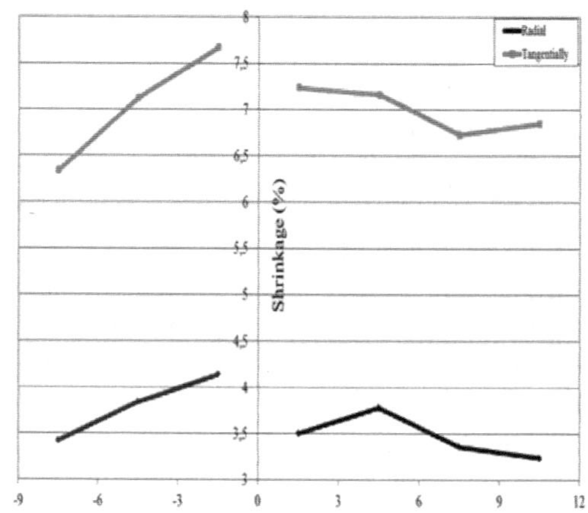

Figure 2c. Block C, distance from pith (cm). Upper tangentially and down radial

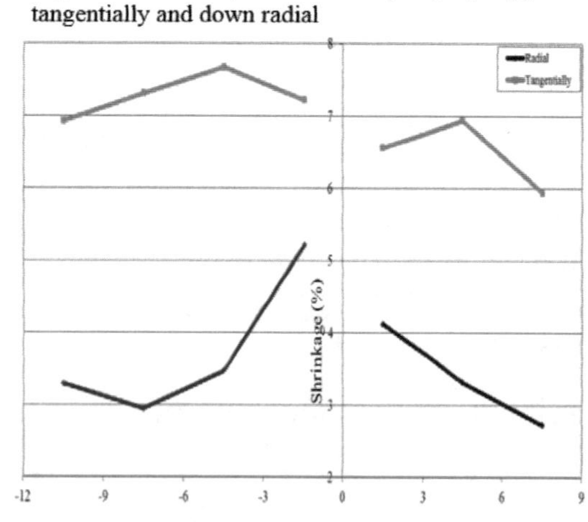

Figure 2 d. Block D, distance from pith (cm). Upper tangentially and down radial

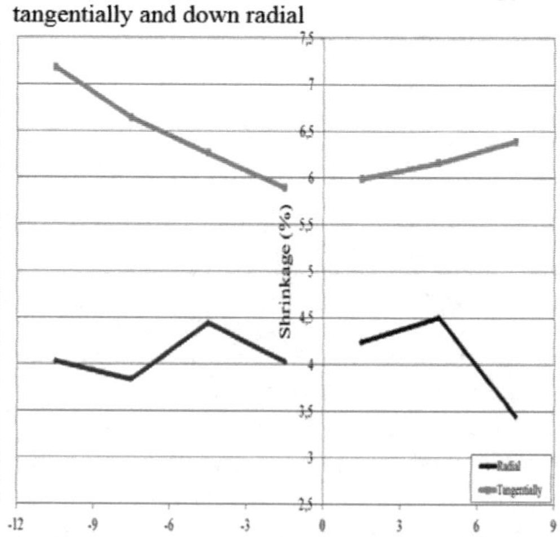

Figure 2e. Block E, distance from pith (cm). Upper tangentially and down radial

Figure 2. Maximum radial and tangential shrinkage changes along with the direction from pith to bark. Upper tangentially and down radial (air dried).

IOP Conf. Series: Materials Science and Engineering **908** (2020) 012004 doi:10.1088/1757-899X/908/1/012004

The effect of hot water extraction (6 and 12 days) on density, shrinkage and fiber saturation point is shown in Table 2 and Figure 3.

Table 2. Density, shrinkage and fiber saturation point (FSP) of chestnut wood before and after extraction (6 and 12 days) with hot water density, radial/tangential.

Sample	Density, g/cm^3	Shrinkage (%)*		Shrinkage (%)**		FSP (%)***	
		Rad.	Tang.	Rad.	Tang.	Rad.	Tang.
Before hot water extraction							
$A_{(1-3, 7-9)}$	0,583	2,54	3,91	4,37	6,58	26,590	27,081
$B_{(1-3, 7-9)}$	0,548	2,22	3,75	3,76	6,27	26,171	26,637
$C_{(1-4, 9-12)}$	0,537	2,10	4,43	3,41	7,03	29,069	29,982
$D_{(19-22, 27-30)}$	0,626	2,21	4,45	3,73	7,26	27,984	29,118
$E_{(19-22, 27-30)}$	0,585	2,41	4,07	4,10	6,48	28,249	30,812
Average	0,576	2,30	4,12	3,87	6,72	27,613	28,726
After 6 days hot water extraction							
$A_{(1-3, 7-9)}$	0,518	3,62	5,68	4,84	7,44	32,701	34,981
$B_{(1-3, 7-9)}$	0,502	3,10	5,69	4,28	7,49	29,206	33,453
$C_{(1-4, 9-12)}$	0,485	2,68	5,79	3,61	7,63	33,437	34,873
$D_{(19-22, 27-30)}$	0,516	3,21	6,46	4,34	8,25	31,140	39,021
$E_{(19-22, 27-30)}$	0,567	3,32	5,29	4,59	6,97	30,517	34,654
Average	0,518	3,19	5,78	4,33	7,56	31,400	35,396
Beforehotwaterextraction							
$A_{(4-6, 10-12)}$	0,584	2,60	4,01	4,41	6,53	26,762	28,728
$B_{(4-6, 10-12)}$	0,550	2,19	3,76	3,71	6,35	26,161	26,283
$C_{(5-8, 13-16)}$	0,540	2,04	4,43	3,37	6,87	28,478	31,302
$D_{(23-26, 31-34)}$	0,624	2,13	4,36	3,83	7,21	25,842	28,712
$E_{(23-26, 31-34)}$	0,579	2,45	3,96	4,19	6,44	27,921	29,730
Average	0,575	2,28	4,10	3,90	6,68	27,033	28,951
After 12 days hot water extraction							
$A_{(4-6, 10-12)}$	0,496	5,96	8,20	7,12	10,02	47,830	42,413
$B_{(4-6, 10-12)}$	0,488	4,71	8,06	5,93	9,78	37,172	43,246
$C_{(5-8, 13-16)}$	0,472	4,47	8,62	5,51	10,10	41,345	72,895
$D_{(23-26, 31-34)}$	0,490	4,53	9,29	5,55	10,91	55,311	50,786
$E_{(23-26, 31-34)}$	0,538	3,14	8,81	4,33	10,65	28,498	45,332
Average	0,497	4,56	8,60	5,69	10,29	42,031	50,934

* Shrinkage from green (wet) to air-dry conditions
** Maximum shrinkage to oven-fry conditions (moisture content 0%)
*** Determination from corresponding dimensional changes (radially, tangentially)

The prolonged extraction duration of the samples led to an average extract of 10.5% and 17.20% for extraction 6 and 12 days, respectively, compared to an average of 19.22% extracts determined according to ASTM standards. Hot water extraction led to 10.07% and 13.57% decrease in density after extraction for 6 and 12 days, respectively. Air-dry shrinkage increased 38.70% radially and 40.29% tangentially after 6 days extraction. This increase was higher (almost double) after 12 days extraction (100% and 109.76% radial and tangential respectively). Maximum shrinkage increased 11.89% radially and 12.50% tangentially after 6 days extraction. This increase was, also, higher after 12 days extraction, radially (45.90%) and tangentially (54.04%). In all cases, fiber saturation point was increased after extraction. Higher increase was found after 12 days extraction. In those cases where lower shrinkage values were found around pith area, they could be attributed to the presence of "juvenile" wood, due to increase of microfibrils angle associated with the presence of such wood or to a higher content of extracts in heartwood region around pith.

Lower radial and tangential shrinkage values in "juvenile" wood (1.5% and 2.8% respectively) compared to corresponding "typical" wood (2.1% and 3.7%) are also reported by Militz et al. [10] for coppice chestnut wood.

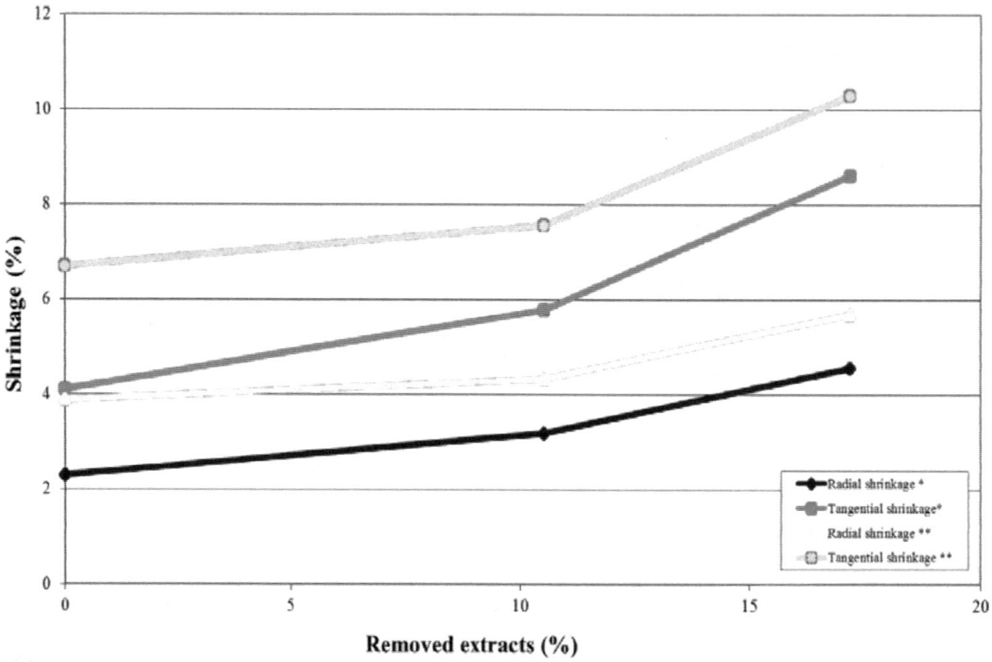

Figure 3. Radial and tangential chestnut wood shrinkage before and after extraction (6 and 12 days) with hot water. * Shrinkage from green (wet) to air-dry conditions ** Maximum shrinkage to oven-fry conditions (moisture content 0%).

4. Conclusions

Mean maximum radial shrinkage of chestnut wood was 3.83% and ranged from 3.58% - 4.12%, while maximum tangential shrinkage was 6.58% and ranged from 6.21% -7.01%. These values, combined with the rest of literature, tend to precisely reflect the shrinkage of chestnut wood from Mediterranean coppice forests. Removal of water-soluble extracts results in an increase in radial and tangential shrinkage. From pith to bark direction, shrinkage seems to have some fluctuations (mainly reductive for radial shrinkage) but not to a large extent. Fiber saturation point calculated from changes in the tangential dimensions was found to be slightly higher than in radial dimensions. This difference led to first changing the tangential dimensions of the wood and then the radial ones, when reducing moisture content, but is, also, a very low difference and does not affect the commercial use of wood.

5. References

[1] Tsoumis G 1994*Wood Science and Technology: Volume A Structure and Properties*(Publications Office of the Aristotle University of Thessaloniki, Thessaloniki, Greece)

[2] Kakavas K et al 2018 Chemical Properties of Greek Stump Chestnut (*Castanea sativa* Mill.)*Nat Prod Chem Res***6**:4 DOI: 10.4172/2329-6836.1000331

[3] Koukos P 1995 Characterization of chestnut timber for industrial uses*Final Technical Report, Contract No MA2B-CT92-002*Task D(EC Forest Project: Innovation in the Chestnut Timber Processing, Spain)

[4] Koukos PK 1997 Some physical properties of sweet chestnut wood grown in Greece*HolzRoh – Werkstoff***55**pp127-129

[5] Voulgaridis E1987 Adsorption - absorption and dimensional stability studies of wood of thirty Greek forest trees. *Scientific Yearbook of the Department of Forestry and Natural Environment, A.U.Th.* Vol. **Λ/2** pp. 63-131

[6] Berti S et al 1991 Per unamiglioreutilizzazione del legnoritraibile dal boscoceduo*Quaderni dell' Istituto per la ricercasullegno* no. **86** (Firenze)

[7] Rousodimos G and Paraskevopoulou A 1996 Chestnut wood (*Castanea sativa* Mill.)*Scientific Yearbook of the Department of Forestry and Natural Environment, A.U.Th.*Vol. **ΛΘ / 1** pp 150-164

[8] Kyuchukov G and Kyuchukov B 1999Physicomechanical characteristics of the wood of *Castanea sativa* Mill. In Bulgaria *XIII Konferencja Naukowa Wydzialu Technologii Drewna SGGW*(WARSZAWA 16-18 LISTOPADA 1999)

[9] Fioravanti M1995 Nature and occurrence of junevile wood in chestnut (*Castanea sativa* Mill.) stems from coppice forest. *ForêtMéditerranéenne***XVI** no.1

[10] Militz H, Busetto D and Hapla F 2003 Investigation on natural durability and sorption properties of Italian Chestnut (*Castanea sativa* Mill.) from coppice stands. *HolzRoh-Werkstoff***61** pp 133-141

IOP Conf. Series: Materials Science and Engineering **908** (2020) 012005 doi:10.1088/1757-899X/908/1/012005

Deposition of Multilayer Films of ZnO by Sol-gel Process on Stainless Steel Substrates for Energy Harvesting Devices

L A Uscanga-González[1], E A Elvira-Hernández[2], R Pérez-Cuapio[3], A Pacio[4], F López-Huerta[5,7]*, E J Alvarez-Sanchez[1,6] and A L Herrera-May[2,7]

1 Doctorado en Ingeniería, Facultad de Ingeniería Mecánica y Eléctrica, Universidad Veracruzana, Gonzalo Aguirre Beltrán s/n, Zona Universitaria, Xalapa, Veracruz, 91090, México
2 Micro and Nanotechnology Research Center, Universidad Veracruzana, Calzada Ruiz Cortines 455, Boca del Río, Veracruz 94294, Mexico
3 Benemérita Universidad Autónoma de Puebla, Facultad de Ingeniería Química, Av. Sn. Claudio y Blvd. 18 sur, Jardines de San Manuel, 72570, Puebla, México
4 Instituto Politécnico Nacional, Laboratorio de Tecnologías y Materiales para Energía, Salud y Medio Ambiente (GESMAT), CICATA Altamira, Km 14.5 Carretera Tampico-Puerto Industrial Altamira, Altamira, Tamaulipas, 89600, México
5 Facultad de Ingeniería Eléctrica y Electrónica, Universidad Veracruzana, Calzada Ruiz Cortines 455, Boca del Río, Veracruz 94294, Mexico
6 Facultad de Ingeniería Mecánica y Eléctrica, Universidad Veracruzana, Gonzalo Aguirre Beltrán s/n, Zona Universitaria, Xalapa, Veracruz, 91090, México
7 Maestría en Ingeniería Aplicada, Facultad de Ingeniería de la Construcción y el Hábitat, Universidad Veracruzana, Calzada Ruíz Cortines 455, Boca del Río, Veracruz 94294, México.
E-mail: frlopez@uv.mx

Abstract. The mechanical vibrations surrounding the environment can be converted in electrical energy by piezoelectric energy harvesters (PEH). The increase on the availability of Wireless Sensor Networks (WSN) increases the need for power supply that replaces ordinary batteries. In this paper, an electromechanical modelling and a deposition of multilayer zinc oxide (ZnO) films for an PEH are presented. The aim of the study is to obtain the thickness of the multilayer ZnO films in different conditions, in order to improve the energy harvesting capacity for a PEH device. The ZnO synthesis was performed by the sol-gel method with dehydrated zinc acetate as a precursor, making deposits of 5, 10 and 15 layers. The depositions were made at room temperature by spin coating at 1440 revolutions per minute for 16 s. An UV-Vis test showed that ZnO were present, with a percentage of peak reflectance at 370 nm and a XRD test showed a preferential crystalline orientation at (002). Also, a Finite Element Modelling (FEM) simulation of the substrate behaviour was performed, functioning as a cantilever beam. When adding a seismic mass and oscillating in the resonance frequency of 103.31 Hz and with 15 layers of ZnO deposited, 2.67 volts were obtained.

1. Introduction

Nowadays, the power generation by fossil resources is the main cause of the climate change, showing an ecological footprint on the flora and fauna [1]. The current energetic demands appear to keep increasing, among various causes, by the rise of portable electronic devices. An Irish research group has reported that in the near future there will be 25 billion electronic devices constantly connected [2]. This increase is boosted by the concept of the Internet of Things (IoT), a trend in the current technology development. The IoT is expected to provide information from the surrounding environment in real time, using different elements interconnected with each other [2]. The micro and nanoelectromechanical systems (MEMS/NEMS) provides technology for the IoT network, like wireless sensors networks (WSN), actuators, data transfer devices, as well as the possibility of a ubiquitous installation and small size [3–5]. The IoT systems are used in different areas, which increase the need of mobile, wireless energy sources, commonly batteries are being used as energy storage device. Existing batteries have been studied for years improving their performance, nevertheless, they present deficiencies. For example, the dispositive portability has increasingly prioritized the design of electronic devices with the smallest dimensions and the least possible weight [6], both affected by the characteristics of the current batteries. Also, in some applications such as sensors for structural monitoring in remote locations, battery charging, or replacement could be costly or infeasible. Furthermore, an improper confinement of discarded batteries is a risk due to chemical substances such as mercury, cadmium, lithium or lead with which they are manufactured. For example, button cells' materials could be toxic to the flora and fauna of the planet, polluting the water. Advances in MEMS technology have allowed the production of small electric power generators as an option for energy supply. The energy produced by human activity like industrial production, the use of automobiles, bridges and buildings, among others, could be used as an energy source to generate electric power [7]. This energy sources are such as: mechanical vibrations, solar power, electromagnetic fields or thermal gradients. Among these energy options, sensitive mechanical vibrations are simple to use and can be exploited by energy harvesters (EH) [8]. The mechanical energy exists in most surroundings as consequence of human activities or natural resources, like breeze of the wind [9]. In addition to the ubiquity of this energy, it is possibility of design and manufacturing EH in the order of micrometers [10], which makes possibly to apply them with biocompatible portable sensors or energize sensors in bridges and buildings [8, 11].

Among the different type of EH, the Piezoelectric Energy Harvesters (PEH) generates relatively high-power density and voltage, which depends on the piezoelectric parameters of each different types of piezoelectric materials available. The principal section about the piezoelectric energy harvesters (PEH) is the piezoelectric material, a type of solid able to gather electric charge as result of applied mechanical stress, which is known as the piezoelectric effect [12]. In the PEH, the kinect energy could be provided from the mechanical vibration over the environment, which causes the desired stress on a determined object and generates electrical power. The zinc oxide (ZnO) is a material with piezoelectric properties that has been the subject of various investigations in recent years due to its physical properties [13]. It is a versatile material and can be deposited on a substrate using different techniques, for example the sol-gel method [14]. The application of different designs to supply energy to WSN are relevant duo the advantage of providing a mobile source of clean energy, with minimal maintenance and exploit the ubiquitous mechanical energy [15]. Hence, in this work, we present the procedure to obtain a multilayer ZnO film with a sol-gel synthesis method by variating the number of layers. Furthermore, a finite element modelling (FEM) simulation was performed to obtain an approximation of the energy that the obtained film could generate. A seismic mass was added to increase the obtained power [16] and decreased the resonant frequency of the PEH.

2. Methodology

The ZnO films were deposited by the methodology proposed by [17]. To synthesize ZnO solution, a Zinc acetate ehydrate was used as precursor, monoethanolamine (MEA) as stabilizer and 2-metoxyethanol as solvent. 14 ml of 2- methoxyethanol were put into a ball flask, and 430 μl of MEA were added and then this mixture was stirred for several minutes at room temperature. After that, 1.55 g of zinc acetate dihydrate was added to the previous solution, and stirred at room temperature for 30 min. The ZnO films was deposited on ANSI 302 stainless steel (SS302) substrates using the spin coating technique at 1440 revolution per minute (rpm) for 16 s. After each coating, the films were preheated in an oven at 200 °C for 5 min. Each deposition was repeated after 5, 10 and 15 times to increase the thickness. Next, the samples were annealed in a muffle furnace at 300 °C for 2 hours under oxygen atmosphere. Finally, a Finite Element Simulation was performed in order to obtain the amount of energy that this film would generate as a EH device.

3. Results and discussion

a) b) c) d) e)

Figure 1. Substrates of SS302 with 5 layers (a), 10 layers (b and c) and 15 layers (d and e). The substrates of 1b and 1d were subjected to annealed at x|300 ° C for two hours.

The ZnO films was deposited on 2x2 cm SS302 steel substrates: one with 5 layers (Figure 1a), two with 10 layers (Fig 1b and 1c) and two more with 15 layers (Fig. 1d and 1e). A 10-layer substrate (Fig 1b) and a 15-layer substrate (Fig 1d) were annealed at 300 ° C for two hours. To measure the thickness of the films, a step was made in each, using 1 molar hydrochloric acid (HCl) for 10 seconds. The films showed qualitative uniformity and adhesion, and in order to corroborate the presence of ZnO in the substrates of Figure 1, characterizations of percentage reflectance were performed. The reflectance percentage results (Figure 2) shows a first peak that starts at 370 nm. This confirms the presence of ZnO because it absorbs light with wavelength around 375 nm.

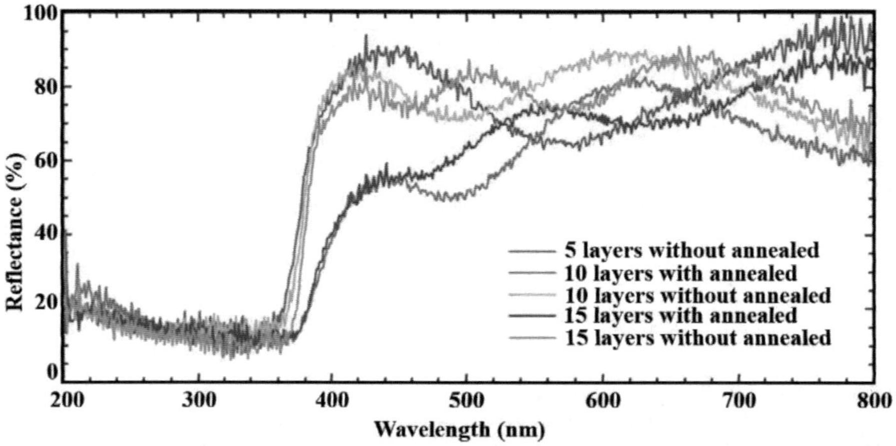

Figure 2. Results of reflectance percentage made to deposited films.

Also, figure 2 shows a sinusoidal oscillation in all measurements after the first crest, which indicates homogeneity in the films. In addition, it is observed that the measurements of the heat-treated films (Figure 1b and 1c) shows a shift to the right, indicating larger grain size in the film.

Table 1. Thickness of the different layers.

Deposition	Thickness (nm)
5 layers without annealed	972
10 layers with annealed	323
10 layers without annealed	593
15 layers with annealed	944
15 layers without annealed	1634

Table 1 shows the thickness of the deposited layers obtained using a profilometer, showing that the thinnest layer was 972 nm and the thickest layer was 1634 nm. Figure 3 shows the results of the X-ray diffraction of the sample with 15 layers, observing peaks with 2Θ values at 25.29 °and 38.55°, it was possible to confirm the presence of ZnO with preferential orientation in the plane (002), as well as the stainless steel that acts as a substrate. Using a FEM software, the data obtained with the subtracted with 15 layers of ZnO was entered to obtain the voltage that would be generated by using it as EH. The SS302 substrate was drawn, posing as a cantilever beam. A seismic mass of neodymium was added with measures of 2x0.5x0.05 cm, this to decrease the resonance frequency of the device.

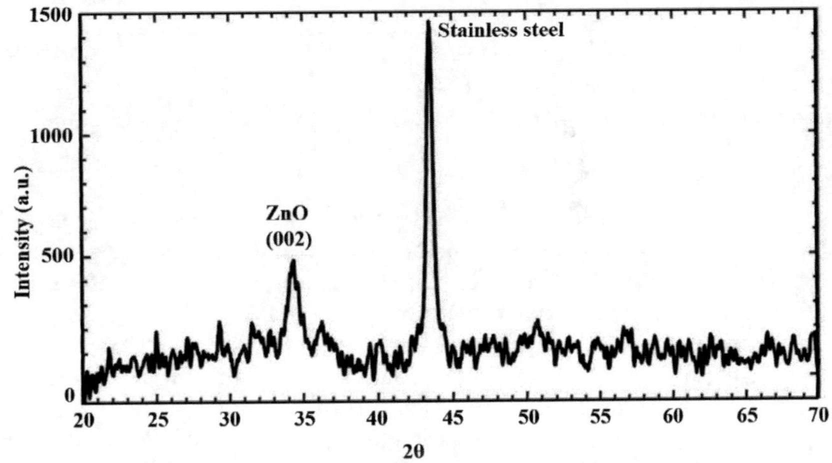

Figure 3. X-ray diffraction of the sample with 15 layers.

To solve the mathematical model, a mesh with 15090 elements was generated, having a greater refinement at the opposite end to the seismic mass as can be seen in Figure 4a. This is because is the area with the maximum strain when vibrations affect the device. Figure 4b shows the behavior of the stresses along the substrate in the first vibration mode, which is at 103.31 Hz.

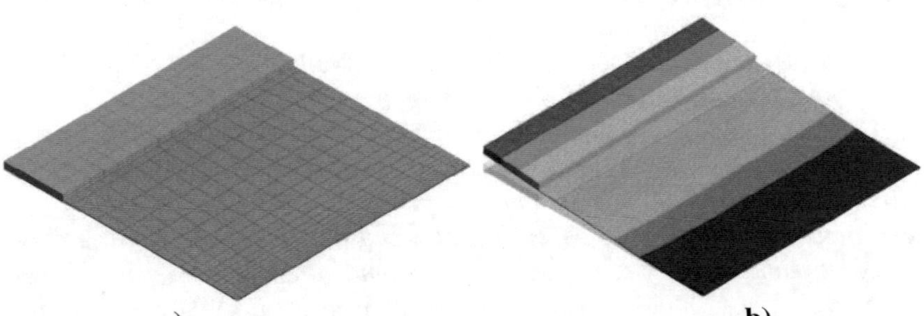

Figure 4. Figure 4a shows the mesh made to the geometry of the substrate with multilayer ZnO film. In the figure 4b shows the strain along the substrate when the device is excited at 103.31 Hz.

Figure 5 shows the graph that describes the voltage generated by the simulated EH, showing a sweep from 103 to 103.5 Hz, passing through the resonant frequency, where the highest voltage value is shown, this being 2.67 volts. As future work will be conducted electrical characterization to determine the voltage and power generated can obtain this device.

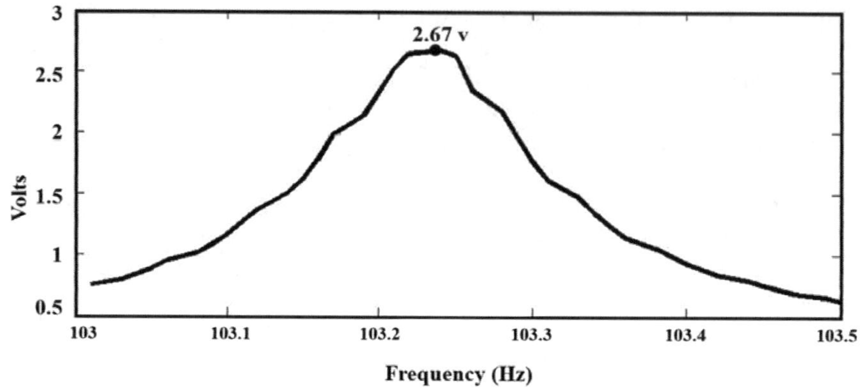

Figure 5. Graph of the frequency sweep to observe the generated voltage.

4. Conclusions

The depositions of ZnO films obtained on SS302 substrates were made using sol-gel method. With the spin coating technique, the number of deposited layers of 5, 10 and 15 layers was varied to increase the thickness of the films. The results of the XRD diffractogram and UV-Vis reflectance confirm the presence of ZnO, with homogeneous films in the preferential plane (002). These films can be used as energy harvesting microgenerators. The simulation performed shows that an EH with a seismic mass of neodymium with measurements of 2x0.5x0.05 cm could generate a potential difference of 2.67 volts when entering its first resonance mode at 103.31 Hz, thus functioning reusing the mechanical energy, like mechanical vibrations, in the environment surrounding the device. As a future work, the fabrication and characterization of the presented PEH will be performed, in order to compare the result obtained in the FEM simulation.

5. References

[1] Bello M O, Solarin S A, and Yen Y Y 2018 The impact of electricity consumption on CO2 emission, carbon footprint, water footprint and ecological footprint: The role of hydropower in an emerging economy *J. Environ. Manage.* **219** 218–30

[2] Mallick D, Constantinou P, Podder P and Roy S 2017 Multi-frequency MEMS electromagnetic energy harvesting *Sensors Actuators A Phys.* **264** 247–59

[3] Miraz M H, Ali M, Excell P S and Picking R 2015 A review on Internet of Things (IoT), Internet of Everything (IoE) and Internet of Nano Things (IoNT) *2015 Internet Technologies and Applications (ITA)* vol 113 (Wrexham, UK/ IEEE) pp 219–224

[4] Dinis H, Colmiais I and Mendes P M 2017 Extending the Limits of Wireless Power Transfer to Miniaturized Implantable Electronic Devices *Micromachines* **8** 359

[5] Tan Y, Dong Y and Wang X 2017 Review of MEMS Electromagnetic Vibration Energy Harvester *J. Microelectromechanical Syst.* **26** 1–16

[6] Wang X, Niu S, Yi F, Yin Y, Hao C, Dai K, Zhang Y, You Z and Wang Z L 2017 Harvesting Ambient Vibration Energy over a Wide Frequency Range for Self-Powered Electronics *ACS Nano* **11** 1728–35

[7] Priya S, Song H C, Zhou Y, Varghese R, Chopra A, Kim S G, Kanno I, Wu L, Ha D S, Ryu J and Polcawich R G 2017 A Review on Piezoelectric Energy Harvesting: Materials, Methods, and Circuits *Energy Harvest. Syst.* **4** 3–39

[8] Yang Z, Zhou S, ZuJ and Inman D 2018 High-Performance Piezoelectric Energy Harvesters and Their Applications *Joule* **2** 642–97

[9] Umaba M, Nakamachi E and Morita Y 2015 Development of the magnetic force-induced dual vibration energy harvester using a unimorph cantilever *Micro+Nano Materials, Devices, and Syst.* **9668** 96684U

[10] Fernandes E, Martin B, Rua I, Zarabi S, Debéda H, Nairn D, Wei L and Salehian A 2018 Design, fabrication, and testing of a low frequency MEMS piezoelectromagnetic energy harvester *Smart Mater. Struct.* **27** 035017

[11] Kim K, Guo J, Liang Z and Fan D 2018 Artificial Micro/Nanomachines for Bioapplications: Biochemical Delivery and Diagnostic Sensing *Adv. Funct. Mater.* **28** 1–19

[12] Invernizzi F, Dulio S, Patrini M, Guizzetti G and Mustarelli P 2016 Energy harvesting from human motion: Materials and techniques *Chem. Soc. Rev.* **45** 5455–73.

[13] Liu H, Zhong J, Lee C, Lee S W and Lin L 2018 A comprehensive review on piezoelectric energy harvesting technology: Materials, mechanisms, and applications *Appl. Phys. Rev.* **5** 041306

[14] Todaro M T, Guido F, Mastronardi V, Desmaele D, Epifani G, Algieri L and De Vittorio M 2017 Piezoelectric MEMS vibrational energy harvesters: Advances and outlook *Microelectron. Eng.* **183–184** 23–36

[15] Vasistha V 2014 PROVIDING ADDITIONAL SOURCE OF ENERGY IN A VEHICLE USING PIEZOELECTRIC MATERIAL WITH NO RECURRING COST *Int. J. Mech. Eng. & Rob. Res.* **3** 127-128

[16] Yazid E and Nugraha A S 2019 Energy Harvesting From a Crane System by a Piezoelectric Energy Harvester: Two Concentrated Moving Masses Approach *Int. J. Mech. Eng. & Rob. Res.* **8** 940-946

[17] Pérez-Cuapio R, Pacio M, Juarez H, Alvarado J A, Guarneros C, Bueno C and Pacio A 2018 Effect of ZnO Film Thickness on its Optical and Structural Properties *J. Nano Res.* **52** 102–14.

Acknowledgments

This research was funded by project PRODEP "Modelado y Desarrollo de Microsensores CMOS-MEMS para la detección de diabetes" and CONACYT trough Grant 633194 and by Program of Doctorate in Engineering of the Faculty of Mechanical and Electrical Engineering of Universidad Veracruzana, México.

IOP Conf. Series: Materials Science and Engineering **908** (2020) 012006 doi:10.1088/1757-899X/908/1/012006

Smart Materials Vis-à-vis Metamaterials

MALIK Abdul Qadeer Malik[1, a]

1School of Chemical and Materials Engineering, National University of Sciences and Technology, Islamabad, Pakistan
E-mail: [a]aqmalik46@yahoo.com

Abstract. Smart materials are custom-designed materials that have certain properties that can undergo significant changes as guided through external influence, such as pressure, temperature, pH, electric or magnetic fields. One of the most amazing developments in the smart materials is that of metamaterials. The origin of Meta stems from Greek meaning "beyond". Metamaterials have rare properties not found in natural occurring materials such that their properties are based on their structure rather than material of their composition. These materials find applications in many fields like public safety, sensors, battlefield communication, ultrasonic devices, solar power equipments, and remote aerospace applications. A synthetic composite material with a structure such that it possesses properties being unusual in natural materials specially a negative refractive index-these materials scatter electromagnetic wave unlike any known material. Thus the shape of the wave emerging from the metamaterials is entirely different from a conventional wave. Besides having a myriad of applications, these materials have the remarkable ability to bend light which can make airplanes, tanks and submarines invisible resulting in the epitome of stealth capability. Metamaterials now find recognition as an enabling technology being the most exciting research area at the crossroads between photonics and nanoscience. Without any exaggeration, the present era is the century of metamaterials. This paper, inter alia, highlights the significance of metamaterials in general and its stealth capabilities for defence sector in particular.

1. Introduction

Smart materials [1] are custom-designed materials possessing certain properties that can undergo significant changes as guided by external influence such as electric or magnetic fields- one of the most amazing developments in the smart materials is that of metamaterials. The Meta [2] stems from Greek meaning "beyond", metamaterials have rare properties not found in natural occurring materials such that their properties are based on their structure rather than material of their ingredients. These materials find applications [3, 4, 5] in many fields like public safety, battle field communication, ultrasonic devices, solar power equipment, and remote aerospace applications besides stealth capability i.e. a type of military camouflage to make personnel, aircraft, ships, submarines etc. less visible(ideally invisible) to radar and other detection methods.

A synthetic composite material with a structure such that it possesses properties being unusual in natural materials specially a negative refractive index – these materials scatter electromagnetic wave unlike any known material. Thus the shape of the wave emerging from the metamaterials is entirely different from a conventional wave.

2. Smart vis-à-vis Metamaterials-Concepts

Smart materials are [6] intelligent in the sense that these respond to their environment. Emerging concepts of smart materials pertain to piezoceramics, self – repairing airplane etc. besides active

IOP Conf. Series: Materials Science and Engineering **908** (2020) 012006 doi:10.1088/1757-899X/908/1/012006

suppression of submarine noise. Due to the capacity of these materials to reconfigure to match the environment – these are called smart rather active materials. In contrast, metamaterials are man-made materials designed to possess properties not occurring in nature. Composite materials such as metals and plastics are assembled in layers to create these materials. Each layer has different shape, geometry, size and orientation of atoms to produce unusual electromagnetic properties such as negative index of refraction. The material with a negative angle of refraction changes the angle of refraction of incident light causing its propagation direction to change to absorb electromagnetic waves thus producing a weaker radar signal. This results in cloaking technology that fully bends all light incident to the targeted area ultimately becoming undetectable to the radar. Thus metamaterials could make radar obsolete to make the airplane invisible.

3. Piezoelectric Ceramics

The piezo- in Greek means to 'press' or 'squeeze'. Piezoelectric sources are versatile devices that measure various processes. Lead zirconate and lead titanate are the basic ingredients of there materials. Their main advantage is that their properties can be adjusted to suit specific applications by using the appropriate zirconate-titanate ratio. These are hard, chemically inert and thoroughly insensitive to environmental conditions. Their mechanical properties resemble ceramic insulators and are synthesized almost by the same processes. Piezoelectrics are ideal for all kinds of electro mechanical transducers. These materials are marketed under the trade name PZT (Lead ZinconateTitanate) Major applications include generators, sonic and ultrasonic transducers, sensors and actuators. [7]
Piezoelectric effect was discovered by Piere Curie in 1880 but it was only in 1950 that manufacture of sensors started using the piezoelectric effect as sensing applications. These materials have very high modulus of elasticity comparable to that of metals and goes up to 106N/m2. Some sensors have working range up to 1000°C. Piezoelectric sensors are used to monitor combustion in IC engines. A major application of piezoelectric transductions is in the defence industry. They are used in the nose fuze of rocket propelled grenade (RPG)-an antitank weapon wherein these transducers produce very high voltage to initiate the base fuze of RPG. [8] When an RPG hits a target, the nose fuze equipped with piezoceramic is crushed and consequently very high voltage is generated and transmitted for the ignition of the base fuze of the rocket to detonate the main high explosive charge which in consequence produces a hyper velocity metallic jet to defeat the armour or any other target fired upon Fig. 1.

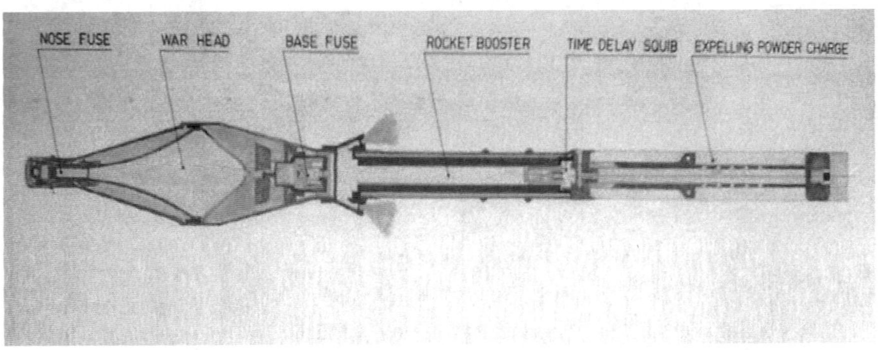

Figure 1. Rocket propelled grenade-an antitank weapon

4. Active Coatings for Signature Suppression

Even modern submarines have acoustic signatures produced by the propulsion system, internal rotating machinery/pumps and the flow of water over/around the hull. The localized excitation of the acoustic modes of the hull effectively mix with the surrounding water and are readily detectable at long range. The role of smart materials here is an active coating; it is possible to measure the amplitude and frequency of an acoustic mode in the hull. If acoustic actuators are then built into the coating, it becomes possible to create 180° out-of-phase acoustic signals of an amplitude and

IOP Conf. Series: Materials Science and Engineering **908** (2020) 012006 doi:10.1088/1757-899X/908/1/012006

frequency that match that of the acoustic mode. The outcome results in destructive interference and a greatly reduced signal radiated into the surrounding. This 'active stealth' approach promises to greatly reduce acoustic signatures by 60 dB.[8]Besides this, R&D is being conducted for the development of advanced materials and coatings through the fabrication of Nano-and ultrafine materials. Nickel aluminum bronze (NAB) thus developed is employed widely in a variety of sea water systems due to its high corrosion resistance [9].

5. Metamaterials: A Novel Approach

A Metamaterial is designed to have a property not found in naturally occurring materials. These are made from layers of several ingredients like metals and plastics and are arranged in repeating sequence being smaller than the wavelength of the phenomena they affect. Their properties depend upon their new structures rather than the ingredients from which there are synthesized. Their precise shape, geometry, size, orientation and arrangement determine their smart unique properties for manipulating electromagnetic waves by blocking, absorbing, enhancing, or bending waves to achieve rare benefits not found in conventional materials.

Metamaterials by directing and controlling the propagation and transmission of specified parts of the light spectrum demonstrate to render an object seemingly invisible. Metamaterial cloaking, based on transformation optics, explains the shielding of an object from view by controlling electromagnetic radiation. Despite the presence of objects in defined locations, the incident waves are guided such that these are not affected by the objects.

Working of conventional lens, Fig. 2 owes to its specific shape. The light converges due to material property of glass having variable thickness. The light ray slows down in the middle of the lens and travels faster near the edges due to less thickness as depicted in Fig. 3. This convergence results into focusing of image. The lens shape must be well-controlled and precise Fig. 4. Lens with a simpler shape can perform the required task with more flexibility in material property such that the light waves can be controlled by controlling material properties which is the basic idea of metamaterials [10].

More material property at the centre and less at the corners can reproduce the similar results of controlling the light wave. The wavelength and amplitude are important characteristics of waves. Creating an array/grid of metamaterial structure with different material in each square having size comparable to or smaller than the wavelength produces a completely arbitrary complex wave front instead of a converging wave shape produced from a conventional lens, Fig. 5. Thus, by creating a powerful and complex wave processing, the stealth capability is achievable.

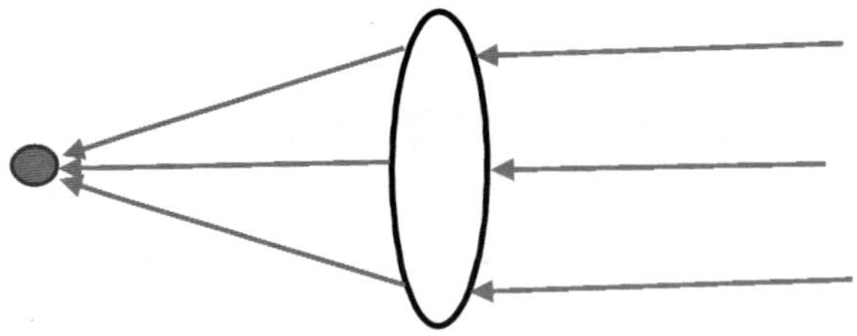

Figure 2. A conventional lens

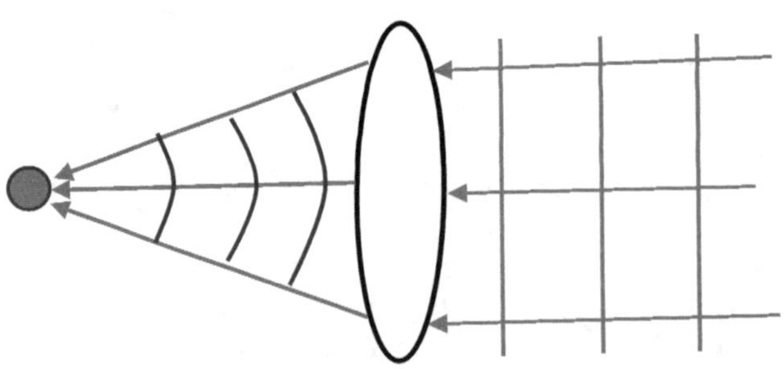

Figure 3. Working of a conventional lens

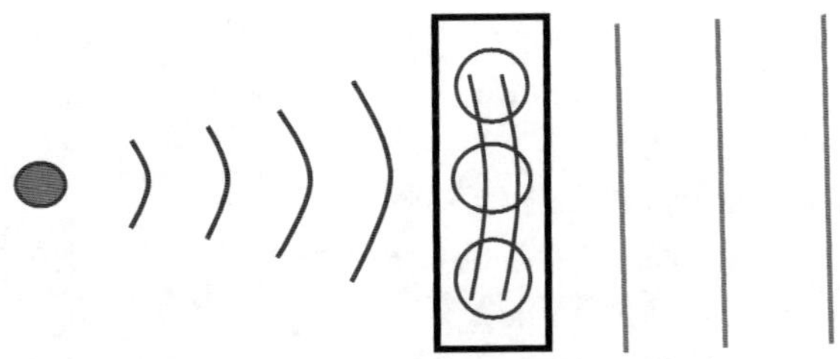

Figure 4. Controlled and precise lens and its working

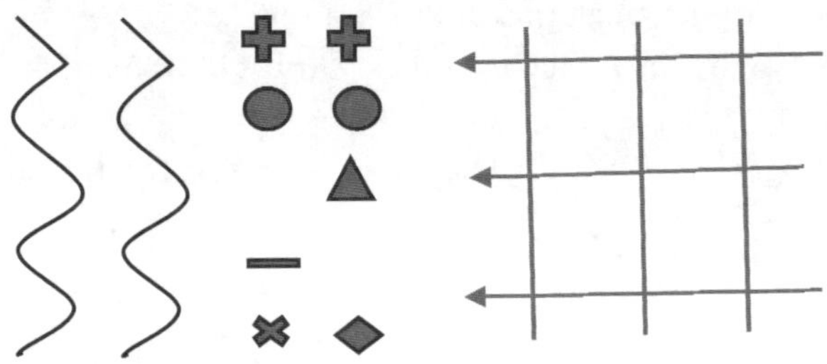

Figure 5. Metamaterials lens producing a complex wave front

6. Century of Metamaterials

It was nigh improbable to predict towards the end of 20th century that the well-established field of electromagnetism would be jolted beyond imagination. This happened in the year 2000 at the start of the new century in the form of a revolution echoed by one word – metamaterials. Ever since the advancement in research, metamaterials have now grown as a full-established discipline termed as the "The century of metamaterials". [11] Accordingly, these metamaterials are artificial materials not natural. [12] The field is a synergic research outcome achieved by the top talent from microwave and optical sciences, solid state physics, material science, acoustics, mechanics, nanotechnology and high performance computing domains.

7. Conclusion

Needless to emphasize that metamaterials have become an extremely exciting research domain. Multiple disciplines are involved in the R&D in this area due to the unique electromagnetic properties provided by metamaterials. Consequently, this will herald the merging of knowledge and expertise across host of areas, in the fantastic advance of metamaterials research. Many amazing and novel concepts and devices such a negative refraction, superlens and invisible cloak have already been witnessed.

Metamaterials now find recognition as an enabling technology and are of the most exciting research area at the crossroads between photonics and nanoscience. Besides having a myramid of exciting applications, the stealth capability is an achievement par excellence with complete flexibility to control the material properties. What could be done next: The sky is the limit! [13]

8. References

[1] World Congress on Chemistry November 15-17,2018 – Rome, Italy

[2] Gurwinder Single, et.al., A review of metamaterials and its application, IJETT-Volume 19, November 6-2015

[3] Practical applications for metamaterials beyond the invisibility cloak – MOUSER Electronics, May:2011
[4] Gurwinder Singh et.al., A Review of Metamaterials& its Application DOI: 10,14445/22315381/IJETT.V19P254
[5] Ten applications of metamaterials IEEE Int. Sym.,2016
[6] Wadley, Haydn, N.G, Characteristics and processing of Smart Materials, AGARD SMP Lecture series, 1996,pp. 1-3
[7] Morgan Technical Ceramics, Piezolectric Ceramics Properties and Applications pp1-12
[8] Wikiviserally, Rocket-propelled grenade (RPG) Para/item 19
[9] Ibrahim, M.E: et.al., Advanced Applications, of Smart Material Research for the Ehancedmentof Australian Defence Capability, DSTO-TN-919, pp. 5-6
[10] Duke University video
[11] Sergei et.al., The century of metamaterials,Journal of optics, Vol 19, Number 8, 2017 (Editorial)
[12] Shining Zhu, Xiang Zhand 2018 Metamaterials: Artificial materials beyond nature, National Science Review **15(2)** 131
[13] Yongmin Liu and Xiang Zhang 20-11 Metamaterials: a new frontier of science and technology, Chem. Soc-Rev; **40**,2494-2507

IOP Conf. Series: Materials Science and Engineering **908** (2020) 012007 doi:10.1088/1757-899X/908/1/012007

Synthesis and Characterization of Thin TiO2 Films Using the Sol-gel dip Coating Method

R M Woo-Garcia[1], L García-González[1], P Argüelles-Lucho[1], C Guarneros-Aguilar[2], B Cervantes[3], A L Herrera-May[1,5] and F López-Huerta[4,5]*

1 Centro de Investigación en Micro y Nanotecnología, Universidad Veracruzana, Veracruz, 94294, Boca del Río, Veracruz, México.
2 CONACYT-Instituto Politécnico Nacional, Materiales y Tecnologías para Energía, Salud y Medio Ambiente (GESMAT), CICATA Altamira, 89600, Tamaulipas, México.
3 Instituto de Neurociencias de Castilla y León, Universidad de Salamanca, Pintor Fernando Gallego, 37007, Salamanca, España.
4 Facultad de Ingeniería Eléctrica y Electrónica, Universidad Veracruzana, Veracruz 94294, Boca del Río, Veracruz, México.
5 Maestría en Ingeniería Aplicada, Facultad de Ingeniería de la Construcción y el Hábitat, Universidad Veracruzana, Calzada Ruíz Cortines 455, Boca del Río, Veracruz 94294, México.
*E-mail: frlopez@uv.mx

Abstract. TiO_2 titanium oxide is a material with many applications, due to its electronic and optical properties, the characterization of the nanostructures developed with this material is carried out in order to determine its physical, chemical, and mechanical properties. This article presents the results of TiO_2 multilayered thin film deposited in Corning glass substrate with the immersion technique using the Sol-gel dip coating method. The quality of the film strongly depends on the internal forces of the polymeric chains which lead to viscosity, gravitational force, surface tension, and gravity as an efficient form of obtaining TiO_2 thin film. The film was submitted to thermal treatment of 400 °C for 2 hours. The results of the experiment show film with a thickness from 139 to 171 nm that features anatase phase, a crystalline quality that corresponds to the number of deposited layers and have a bandgap close to 3.5 eV. The results show that TiO_2 thin film can be used in applications that require high permittivity or high transmittance.

1. Introduction

Titanium oxide (TiO_2) has become a very important material for various fields of investigation, due to properties such as chemical structure and mechanical stability, which is important for rust protection combined with photocatalytic attributes with antimicrobial uses. Its low density and thermal stability, biocompatibility, and physical, optical, and electrical properties also make it an attractive material [1]. TiO_2 has the ability to be transparent in visible light yet absorb ultraviolet light. Its potential applications in fields such as electrochemistry, water decontamination [2] can be used for solar protection, batteries, and optical coatings in the microelectronic industry and for the destruction of aerial contaminants and the elimination of harmful bacteria, as well as carcinogenic cells. It has also been used in diverse environmental applications for the decomposition of unwanted toxic organic compounds [3].
TiO_2 has recently been used for applications such as gas sensors and acidity sensors mainly because of its electrical, morphological, and structural properties, which are of great importance to the crystalline

phase, the size of the grain of crystal, the surface, and if applicable, the type of dopant and its concentration [4]. There are many methods to obtain thin TiO_2 film, such as atomic layer deposition (ALD) [5], chemical steam deposit [6], sol-gel [7,8], pyrolysis spray assisted with ultrasound [9], hydrothermal synthesis, and physical deposit methods [10-12]. In general, of all the wet chemical methods, the sol-gel method is most commonly used for the preparation of TiO_2, it is considered to be one of the simplest and most affordable for the manufacture of the thin film. The sol-gel method is the evolution of an inorganic network through the formation of colloidal suspension (sol) and the gelation of sol to form a network in the continuous liquid phase (gel). The advantage of sol-gel is that it facilitates obtaining TiO_2 crystals of a nanometric size of high purity, offering better control of structure growth. Moreover, the porosity and particle size can be manipulated at atmospheric pressure and at a relatively low temperature [13].
The coating of TiO_2 solutions is a liquid depositing process which is obtained using the sol-gel method and generates a complete transformation from sol to gel until it turns into xerogel when the substrate is taken out of the solution bath at a constant speed (U) (see Figure 1). This method produces evaporation and draining of the solution solvent and generates a thin film of TiO_2. It is a dynamic and complex depositing process, frequently used due to its benefits and is a simple film creation method that provides good reproducibility [14]. Figure 1 shows the deposit method when the substrate is abstracted from the solution. However, it is complicated to predict the topography of all of the coating area during depositing by means of immersion, since this is completely dependent on time, the agitation of the solution, the induced evaporation of the of the solvent, the precursor concentration, the temperature of the solution during deposits, the viscosity gradients, and the aging of the solution [15]. The pore size of the TiO_2 film is controlled with capillary pressure, condensation rates, and the evaporation of the solution. The microstructure of the film depends on the structure of the inorganic species of sol, the adding times, the surface tension of the solvent and the capillary forces when the substrate is taken out of the solution [16]. Finally, the thickness of the film and the uniformity as well strongly depend on the flow conditions when the substrate is in the sol bath, the emersion time, and the evaporation of the solution during emersion.

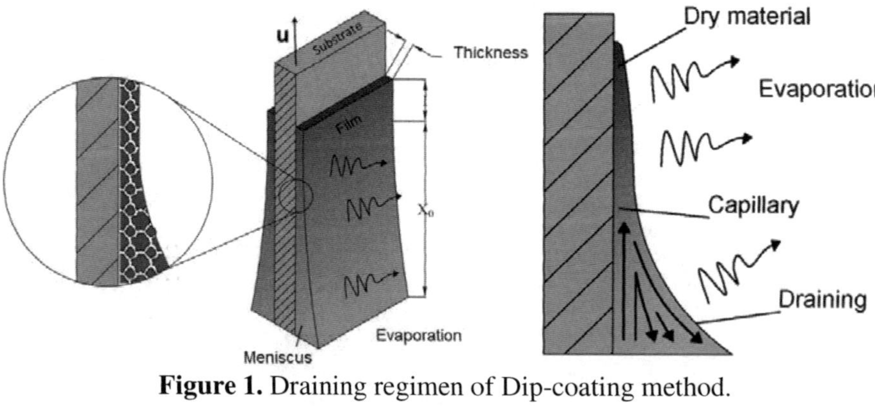

Figure 1. Draining regimen of Dip-coating method.

2. Methodology

A sol solution was made using 7.4 ml of titanium isopropoxide (IV) (TTIP, 97%, Sigma Aldrich), mixing it with 25 ml of ethanol (99.7%). This precursor was cooled with ice until it reached a temperature close to 4 °C, after which it was agitated for 5 minutes. The precursory solution was hydrolyzed with the slow addition of a cold mixture which consisted of 0.625 ml of deionized water and 0.04 ml of achlorhydric acid (36.6%) diluted in 25 ml of ethanol. Acid was added as a system catalyzer. The TiO_2 sol was magnetically agitated for 8 minutes undercooling. The final sol was transparent and in liquid phase. If the mixture is maintained well shut, it remains stable for at least 3 days.
The TiO_2 sol was deposited on commercial corning glass using the dip coating technique, to obtain thin films, an immersion system was used making the deposits at a speed of 2 cm min^{-1}, the substrate was kept for 20 seconds inside the sol for the absorption of the liquid. Different thicknesses of the films were obtained depending on the multilayers deposited, which were from one to five immersions. After each

immersion, the films were removed from the system and annealed at 250 °C for 8 minutes. The annealing was performed immediately, ensuring the densification of the sol and avoiding the condensation of the nanoparticles agglomerated by the reagents. Figure 2 shows the methodology for TiO_2 thin film production.

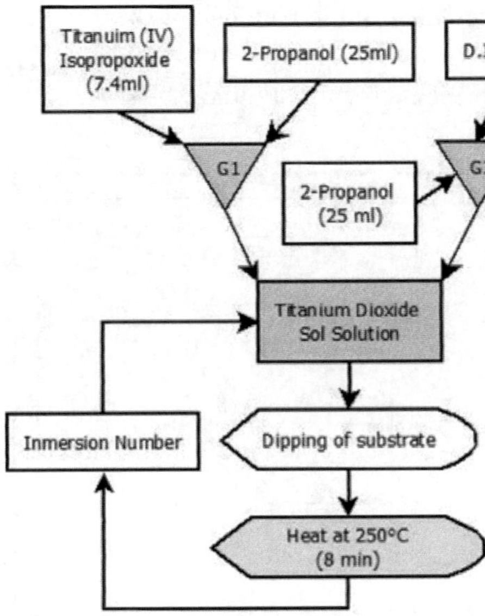

Figure 2. Preparation procedure of TiO_2 thin films synthesis by sol-gel process.

The films were subjected to thermal treatment at 400 °C for 2 hours. The thickness of the film was measured using a Dektak 150 Profilometer, the transmittance and absorbance of the films was measured using a Uv-Vis Evolution 600 Spectrophotometer, the Raman spectra were obtained using Horiba LabRam HR equipment with a 638.8nm helium neon laser (He-Ne).

3. Results and discussion

The film thickness measurement was performed by scanning the surface taking data at the intersection of the substrate and the film every 333 nm over 200 um, Table 1 shows an average of 3 quantifications performed for each film.

Table 1. Average Thickness Titanium Dioxide Films.

Immersion Numbers (Dippers)	Average Thickness (nm)
1	139.23
2	146.45
3	153.32
4	166.17
5	171.75

Figure 3 shows the absorbance spectra, the Tauc plot is used to calculate the bandgap (*Eg*), using the value of the absorption coefficient α given by the Beer-Lambert ratio.

$$\alpha = 2.303 \log\left(\frac{T}{d}\right) \qquad (1)$$

where d is the sample thickness and T is the transmission. Equation 2 represents the ratio of Tauc as a function of $h\nu$ (photon energy), the TiO_2 has an indirect band interval, therefore, $n = ½$ is the transition value that has been used. The values of *Eg* have been estimated by making an extrapolation of the

IOP Conf. Series: Materials Science and Engineering **908** (2020) 012007 doi:10.1088/1757-899X/908/1/012007

absorption slope, acquiring the value of the intersection of this slope with the photon energy when $(\alpha hv)^{\frac{1}{2}}$ is 0.

$$(\alpha hv)^n \propto (hv - E_g) \qquad (2)$$

h is the Planck constant, v is the photon frequency, Eg is the bandgap and n indicates the type of optical transition. It is important to note that the value of the bandgap of the Tauc plot depends on using the appropriate function to match the present electronic transition and the appropriate film thickness [17]. The Tauc plot as $(\alpha hv)^{\frac{1}{2}}$ versus hv is represented within Figure 3. The value of Eg is close to 3.5 eV.

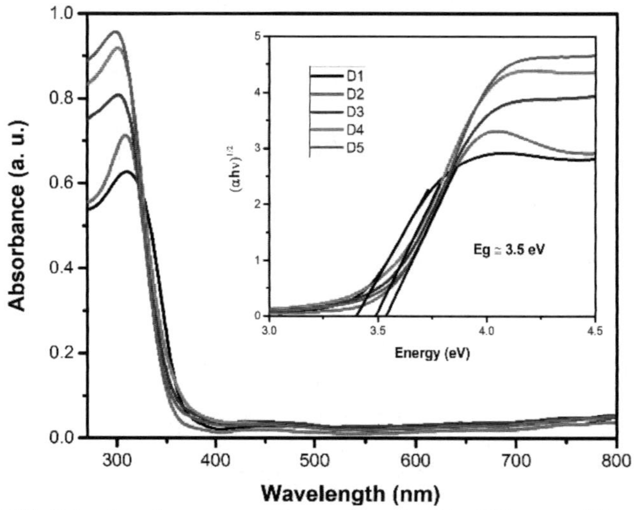

Figure 3. UV-Visible absorbance and Tauc plot of TiO_2 Films at different thickness

A correlation in the increase in the optical bandgap value and the thickness of the film can be observed. Various factors cause Eg displacement towards the right. Different studies show that this phenomenon is largely present due to the Burstein-Moss effect, which describes an increase in the bandgap due to an increase in free carrier concentration [18].

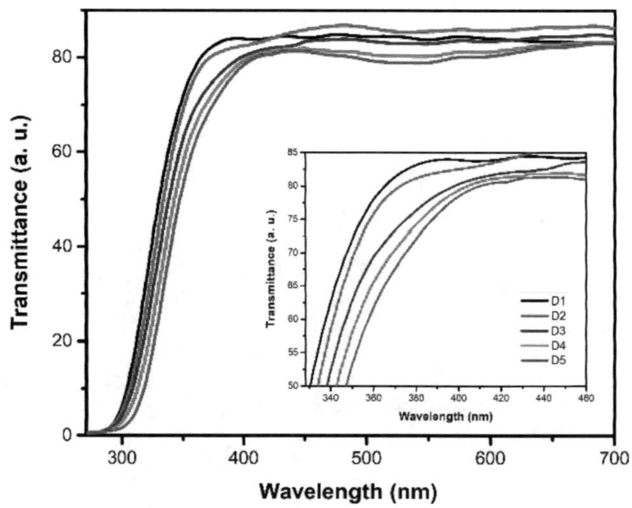

Figure 4. Transmittance spectra of TiO_2 Films.

The transmittance curves of all the film reflect a transmission percentage above 75%, see Fig.4, the spectrums exhibit a perceptible shoulder in the nearby UV region, which is related to the porosity of the film. This property is reduced with an increase in deposited layers.
On the other hand, the first order TiO_2 Raman dispersion in anatase phase resides in the D_{4h} ($I4_1/amd$), spatial group, and according to group analysis, there are six active vibrational Raman modes, positioned at 144cm^{-1} (Eg), 197cm-1 (Eg), 399cm^{-1} (B_{1g}), 513cm^{-1} (A_{1g}), 519cm^{-1} (B_{1g}) and 639cm^{-1} (Eg) [19]. Figure 5 shows the Raman spectrums of the thin TiO_2 films, the spectral characteristics obtained experimentally coincide with the reported anatase phase.

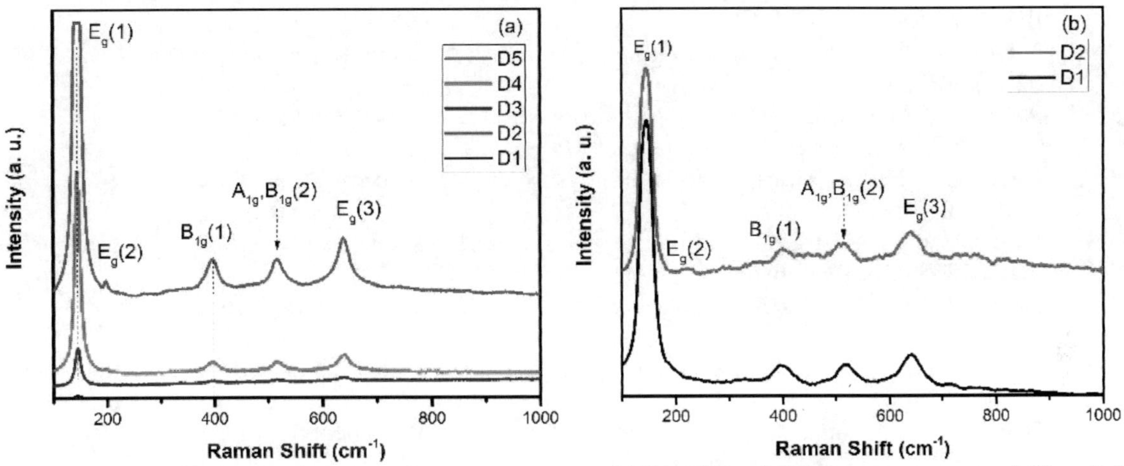

Figure 5. Vibration modes in micro-Raman spectrum of TiO_2 Films (a) Raman sprectrum of films D5-D3 (b) Close up to Raman spectrum of films D1 and D2.

The main differences of the spectrum are related to an increase in the amplitude of vibrational modes $Eg(1)$, $B_{1g}(1)$, and Eg (3) when increasing the number of film layers, in some cases Eg (2) is imperceptible. Anatase mode B_{1g} (1) symbolizes an exclusive vibration of titanium atoms. Peak A_{1g} indicates that there is a vibration of oxygen atoms, unfortunately, this peak overlaps with mode B_{1g} (2) and makes it impossible to indicate whether the A_{1g} + B_{1g} (2) increase is due to possible oxygen vacancies or defects in the titanium-oxygen network of the structure Eg (1), Eg (2), B_{1g} (2), and Eg (3) combine vibrational modes of oxygen and titanium atoms [20].

4. Conclusions

Titanium oxide is an attractive material due to its electrical and optical properties, used in many technological applications, including as a sensor film. Thin TiO2 film was synthesized using the sol-gel method and deposited by coating on Corning glass substrates. Deposits from 1 to 5 layers of different thicknesses were obtained, and once the film reached a xerogel state, they were submitted to a thermal treatment of 400° C for 2 hours. The film had transparency higher than 75% in the visible range of the electromagnetic spectrum that can be related to a high preferential orientation or crystallinity. It was possible to obtain optical parameters, surface attributes, and the 3.5eV bandgap interval at the transmittance and absorbance spectrums. The Raman spectrum indicates an anatase phase that increases with the growth of the film, the vibrational modes indicate that the presence of oxygen drives the current leakage lower, which results in higher resistivity. These phenomena suggest that thin titanium dioxide film made with the sol-gel process acts as an insulator with high permittivity.
The results suggest that the film can be used as a good insulator with high permittivity dielectric as elements for sensitive devices such as a pH sensor or as a gas sensor for an extended gate transistor, moreover, its use in applications that require high transmittance, such as solar cells, could also be considered. In the future, studies about the characterizations of photoluminescence and DRX will be carried out.

IOP Conf. Series: Materials Science and Engineering **908** (2020) 012007 doi:10.1088/1757-899X/908/1/012007

Studies of characterizations of photoluminescence, X-Ray diffraction and hall effect will be conducted in the future.

5. References

[1] Uudeküll, P., Kozlova, J., Mändar, H., Link, J., Sihtmäe, M., Käosaar, S., & Tätte, T. 2017, Journal of Magnetism and Magnetic Materials, **429**, 299-304.

[2] Macwan, D. P., Dave, P. N., & Chturvedi, S. 2011, Journal of materials science, **46(11)**, 3669-86.

[3] Li J-G, Kamiyama H, Wang X-H, Moriyoshi Y, Ishigaki T. 2006, J Eur Ceram Soc 26:423.

[4] Bai, J., & Zhou, B. 2014, Chemical reviews, 114(19), 10131-10176.

[5] Niemelä, J. P., Marin, G., & Karppinen, M. 2017. Semiconductor Science and Technology, 32(9), 093005.

[6] Alotaibi, A. M., Sathasivam, S., Williamson, B. A., Kafizas, A., Sotelo-Vazquez, C., Taylor, A., ... & Parkin, I. P. 2018, Chemistry of Materials, 30(4), 1353-1361.

[7] Nebi, M., & Peker, D. 2016, Journal of physics: conference series (Vol. 766, No. 1, p. 012026). IOP Publishing.

[8] Parthasarathy, P. 2018, IOP conference series: materials science and engineering (Vol. 360, No. 1, p. 012056). IOP Publishing.

[9] Rasoulnezhad, H., Hosseinzadeh, G., & Yekrang, J. 2018. Journal of Nanostructures, 8(3), 251-258.

[10] Zhang, S., Hu, Y., Hu, L., Zhang, Y., Zhou, Y., & Li, B. 2019. Micro & Nano Letters, 14(1), 42-44.

[11] Bhardwaj, S., Basumatary, P., & Agarwal, P. 2018. AIP Conference Proceedings (Vol. 1953, No. 1, p. 100043). AIP Publishing.

[12] Varghese, L., Patra, A., Mishra, B., Khushalani, D., & Gopal, A. V. (2018). Visible absorbing TiO2 thin films by physical deposition methods. arXiv preprint arXiv:1807.11748.

[13] Barbosa, A. I., Borges, J., Meira, D. I., Costa, D., Rodrigues, M. S., Rebelo, R., ... & Reis, R. L. 2019. Materials Science and Engineering: C, 100, 424-32.

[14] Gelover, S., Mondragón, P., & Jiménez, A. 2004. Journal of Photochemistry and Photobiology A: Chemistry, 165(1-3), 241-6.

[15] Faustini, M., Louis, B., Albouy, P. A., Kuemmel, M., & Grosso, D. 2010, The Journal of Physical Chemistry C, 114(17), 7637-45.

[16] Brinker, C. J., & Hurd, A. J. 1994, Journal de Physique III, 4(7), 1231-42.

[17] Shkir, M., Ganesh, V., Yahia, I. S., & AlFaify, S. 2018. Journal of Materials Science: Materials in Electronics, 29(18), 15838-15846.

[18] Pérez-González, M., Tomás, S. A., Santoyo-Salazar, J., & Morales-Luna, M. 2017, Ceramics International, 43(12), 8831-38.

[19] Nezar, S., Saoula, N., Sali, S., Faiz, M., Mekki, M., Laoufi, N. A., & Tabet, N. 2017. Applied Surface Science, 395, 172-179.

[20] https://rruff.info/S.

Acknowledgments

This research was funded by project PRODEP "Modelado y Desarrollo de Microsensores CMOS-MEMS para la detección de diabetes" and CONACYT trough Grant 588062 by Program of Doctorate in Materials and Nanoscience of the Research Center in Micro and Nanotechnology of University of Veracruz (MICRONA-UV), México.

AUTHOR INDEX

Institute of Physics
Dirac House, Temple Back
Bristol BS1 6BE UK

ISSN: 1757-8981
ISBN 978-1-7138-3164-8

CURRAN ASSOCIATES INC.
proceedings.com